The Face-to-Face Principle

Science, Trust, Democracy and the Internet

Harry Collins, Robert Evans, Martin Innes, Eric B Kennedy, Will Mason-Wilkes and John McLevey

Cardiff University Press | Gwasg Prifysgol Caerdydd

Published by
Cardiff University Press
Cardiff University
PO Box 430
1st Floor, 30-36 Newport Road
Cardiff CF24 0DE

https://cardiffuniversitypress.org

First published 2022

Cover design by Hugh Griffiths
Front cover image by Envato

Print and digital versions typeset by Siliconchips Services Ltd.

ISBN (Paperback): 978-1-911653-29-5
ISBN (XML): 978-1-911653-32-5
ISBN (PDF): 978-1-911653-33-2
ISBN (EPUB): 978-1-911653-30-1
ISBN (Mobi): 978-1-911653-31-8

DOI: https://doi.org/10.18573/book7

The full text of this book has been peer-reviewed to ensure high academic standards. For full review policies, see https://www.cardiffuniversitypress.org/site/research-integrity/

Suggested citation: Collins, H., et al. 2022. *The Face-to-Face Principle: Science, Trust, Democracy and the Internet*. Cardiff: Cardiff University Press. DOI: https://doi.org/10.18573/book7. Licence: CC-BY-NC-ND 4.0

To read the free, open access version of this book online, visit https://doi.org/10.18573/book7 or scan this QR code with your mobile device:

Contents

Appendixes 223

Acknowledgements

Mike Gorman and Alun Preece made significant inputs to this project from the beginning to near the end, which influenced it in important ways. Nicky Priaulx was one of inspirers of the project. Darrin Durant offered selfless help in all matters concerning political science, especially the analysis of democracy. Charles Thorpe, Daniel Kennefick, Edgar Whitley, Jeff Shrager and Patrick Dahl provided useful information, ideas and advice. A number of researchers who have looked at remote medical consultations helped Collins with the chapter on that topic that was not included in the end. Without Riccardo Sapienza, Bill Barnes and Willow Leonard-Clarke, the section on scientific conferences would have been much thinner at best. The meetings of Cardiff's Centre for the Study of Knowledge, Expertise and Science (KES), which converted itself during lockdown into an international seminar, provided regular insights and reassurance. Four anonymous referees and a fifth referee, Brian Martin, who refused anonymity, made suggestions that have been very influential. Our copy editor was assiduous and saved us from many mistakes. This book has six authors, and they each owe the other five a big debt of gratitude for making it possible when it would have been all too easy to argue it to a standstill.

The Authors

Harry Collins is Distinguished Research Professor in the School of Social Sciences at Cardiff University and Director of the Centre for the Study of Knowledge, Expertise and Science. He is a Fellow of the British Academy.

Robert Evans is a Professor of Sociology in the School of Social Sciences at Cardiff University. Working in the field of science and technology studies, he is particularly interested in the nature of expertise and the Imitation Game.

Martin Innes is Director of the Crime and Security Research Institute and a Professor in the School of Social Sciences at Cardiff University. He is particularly interested in state-sponsored disinformation and misinformation.

Eric B Kennedy is an Assistant Professor of Disaster and Emergency Management at York University in Toronto, Canada. His research focuses on decision-making, expertise and knowledge systems in crisis contexts.

Will Mason-Wilkes is a Research Fellow in Popular Culture and Media in the Science, Knowledge and Belief in Society Research Group at the University of Birmingham.

John McLevey is an Associate Professor in the Department of Knowledge Integration and Department of Sociology & Legal Studies at the University of Waterloo, Ontario. He works primarily in the areas of computational social science, political sociology, environmental sociology, the sociology of science, and cognitive social science.

The Wide Reach of the Argument

The internet is changing the way knowledge, in the broadest sense, is made and understood. It is a change from making knowledge predominantly in small groups via face-to-face interaction to making knowledge predominantly via remote interaction. We think this change, if it is too radical, could be disastrous for the long-term future of what we call 'pluralist democracy', or 'structured choice democracy', not to mention science and the very idea of truth itself.[1] We would like to stop or slow this change, so it never becomes the accepted standard. This book is meant to explain what the change is, why it should be halted or reversed, and what needs to be done to stop it and its more dangerous consequences.

We started putting together the ideas and gathering the contributors to this book in the summer of 2018. We finished a first draft almost two years later in March 2020. Astonishingly, we found ourselves instantly living through a natural experiment that we might have designed to illustrate the book's thesis. The natural experiment is the coronavirus (COVID-19) pandemic lockdown, which resulted in a massive switch from face-to-face to remote communication.[2] But it meant we had to prepare a rather different second draft which reacted to what we were seeing happening around us.

Two of the triggers for writing the book were the 2016 election of Donald Trump in the USA and the Brexit referendum in the UK, both of which, it seemed to us, had been heavily influenced by the way remote communication could be used and manipulated. Later, by which time we were well into the

---

[1] Pluralist democracy, which is a term that some of us have been using for some years, is a subset of what we will define, in Chapter 11, as 'structured choice democracy'.

[2] Hereafter we will abbreviate 'coronavirus (COVID-19) pandemic lockdown' in the text according to the rhythm of the sentence.

How to cite this book chapter:

Collins, H., et al. 2022. *The Face-to-Face Principle: Science, Trust, Democracy and the Internet.* Pp. 1–20. Cardiff: Cardiff University Press. DOI: https://doi.org/10.18573/book7.a. Licence: CC-BY-NC-ND 4.0

project, the 2019 Brexit-related election reinforced the point still further. And then, still more terrifying, was the build-up to the November 2020 American election, which coincided with the writing of the final draft and reached its climax in the very days that the penultimate draft of the book was being finished. Once more, the book had to change in response to what was happening. The morning of 4 November, when it appeared that Trump would win a second term, was one of the worst in the lives of many of us because we could see *and feel* what the abandonment of the concept of truth would mean.

The book is divided into three parts: Parts I and II, which were informed by the democracy question, now make reference to these changes throughout, but Part III, which is strongly influenced by the 2020 election, has a much stronger focus on immediate events, the short-term and long-term future, and the urgent need for change if pluralist democracy is to last. The cognitive structure of the book is now like an elongated reverse funnel: it starts with a narrow focus covering three chapters on how face-to-face communication works and why it is vital in science; Part II consists of six chapters showing, in various ways, why remote communication cannot replace face to face; Part III is, again, three chapters which widen out from the future of science to the future of democracy. Completing this final draft more than six months later, after the period of normalcy associated with the start of the Biden presidency in the US, we sense that the style of some passages toward the end of Part III might seem overwrought, and less hedged about by qualifications than normal academic treatises. But we have decided to leave these passages as first written: they are a document of those times, even more pertinent since so many of our academic colleagues seem complacent in the light of the way the 2020 election turned out and the subsequent period of calm. But the last people who should be complacent are academics: as we will try to show, Trumpism has not gone away and academics' *raison d'être* – the search for truth – will not survive the coming cultural changes unless the continuing problem is understood and confronted.

To give some sense of the changes in succeeding drafts of the book, the first draft was called 'Face-to-Face: Communication and the Liquidity of Knowledge', and most of what was there remains in Parts I and II, but the new and wider perspective and its portents are reflected in the new title. The changes may have given rise to a more complex thesis but we think that events, however much we regret them, have made its relevance for life in the 21st century even more unarguable. The thesis, as explained, now runs all the way from how scientists communicate to why democracy is dying. To anticipate, the whole thesis is pulled together in Figure 12.2 (see Chapter 12, Non-specialists and MMR) but it is probably necessary to get the sense of the book to understand the figure.

The Face-to-Face Principle

Since the new title refers to the 'Face-to-Face Principle' we should immediately say what it is. Michael Polanyi invented the term 'tacit knowledge' to describe what could be known but not necessarily explicated and that idea, given a sociological gloss, informs much of what is said in this book. The 'Face-to-Face Principle' is similar to Polanyi's claim, on page 7 of 'The Logic of Tacit Inference', that 'all knowledge is either tacit or rooted in tacit knowledge'. The Face-to-Face Principle states:

All human communication is either face to face or rooted in face-to-face communication.

One thing this means is that every kind of remote communication, from smoke signals onwards, depends on the language which makes that which is being communicated comprehensible. This language must be mutually understood at the outset, and fluency in the common language can only have been established by face-to-face interaction.[3] Even more important, remote communication has enormous benefits but, special circumstances aside, only after the face-to-face work of developing the trust which gives the communication its value has been done.

The advantages of the internet

A surprising thing the natural experiment of the pandemic lockdown has revealed is that because of the internet, various aspects of social existence have been far less damaged than they might have been if the same trauma had unfolded twenty or so years ago. For example, though, at the time of writing these sentences (July 2020), face-to-face academic meetings can no longer take place, Zoom and other remote platforms are booming, and by using them, it might seem that academics can go on much as before with little disadvantage, with some even taking the opportunity to argue that the 'success' of this experiment heralds the end of the scientific conference along with its carbon footprint and built-in elitism. There is a small loss in the vitality of interchanges

---

[3] Schutz also distinguishes between direct and indirect experiences, placing them at opposite ends of a continuum and making a similar claim about the foundational value of the former:

> indirect social experiences derive their original validity from the direct mode of apprehension ... There is the whole world of cultural objects, for instance, including everything from artifacts to institutions and conventional ways of doing things ... I can "read" in these cultural objects the subjective experiences of others I do not know. Even here, however, I am making inferences based on my previous direct experience of others. (Schutz 1972, p. 182)

What has changed since Schutz's phenomenological observations, which were first published in the 1930s, is the development of technologies that enable the indirect to mimic the direct.

and some increased fatigue but the ability to talk with people across the globe has been greatly enhanced.[4] On the other hand, the very felt success of the remote in this forced experiment means that a short-term positive experience is encouraging demands to make it permanent; but to abandon the face to face would be disastrous and, in Chapter 10, we will discuss the particular problem for scientific conferences and meetings.

Terminology: the meaning of 'face to face'

When we use the term 'face to face' in this book, or when we use the acronym 'F2F', we mean that the persons involved are physically co-located – sharing buildings, meeting rooms and dining spaces where appropriate. This is the usage found in the dictionary, that we know to be used in respect of medical doctors' consultancy practices now that remote and online consulting is growing apace and personal consulting is becoming hard because of the pandemic, and this is the usage we thought was universal. But we find that in the pandemic, the term 'face to face' is sometimes used to refer to video-platform mediated interaction where individuals can see each other's faces even though they are located remotely from each other. We believe that this usage is mostly unthinking but sometimes it seems to be chosen deliberately by those who want to press the advantages of remote communication – in the way that the philosopher, Charles Stevenson, in his paper written in 1938, called a 'persuasive definition'. Here, we will stick to the traditional use and refer to video interaction as '*imitation* face to face'. The difference we are getting at is something like the difference between referring to surimi as 'crab sticks', on the one hand, or 'imitation crab' on the other. To contrast with the acronym F2F, we will refer to all forms of remote communication as R2R, noting that internet-mediated remote communication is sometimes referred to elsewhere as screen to screen (S2S), or computer mediated communication (CMC).

Knowledge and communication

We argue that face-to-face interaction is essential for the creation of sound knowledge and that too much remote interaction is dangerous to the creation of sound knowledge, at least if the substitution of one for the other goes too far and for too long. The deepest problems are long term. Since we live in the short term this long-term danger cannot be empirically illustrated until it is likely to be too late to reverse it, so our argument is, in this sense, 'theoretical'. Nevertheless, we try to base it on relevant case studies and a variety of related

---

[4] Though there is already accumulating evidence that all is not well. For example, women are submitting fewer papers and uploading fewer preprints, and men submitting and uploading more (Collins 2020).

research, much of which focuses on remote and face-to-face communication in scientific and technical fields; the argument works outwards, from a narrow but deep empirical base to a conclusion that affects social and political existence writ large.

Because of the scale and reach of the general conclusion, many will be able to cite counter-examples and exceptions from their own experience or their own scholarly reading: lots of potential readers of this book will have experienced face-to-face interactions where they were misled and will know that confidence tricksters, criminals in general, malign cultists and non-benevolent dictators depend on face-to-face interaction as much as well-intentioned scientists rely on it, so they will know that F2F is not always good. Many readers will also have experienced at least some sound and productive remote interactions with people who they have never met, and they will be able to cite published studies of successful and sound innovation managed remotely, so they will know that R2R is not always bad. Indeed, we will, ourselves, supply examples counter to the drift of our main thesis both from our own experience and from our own encounters with the literature; as we will describe, we ourselves are avid users of remote communication and could not do our work without it. This is to re-emphasise that we are not saying that *all* F2F is good and *all* R2R is bad. What we *are* saying is that face-to-face communication is a necessary but not sufficient condition for the creation of sound knowledge *in a society*; we are not saying that F2F is a necessary condition for every *individual* instance of the creation of sound knowledge. What we *are* saying is that a society that takes the remote, not the face to face, to be its foundational mode of communication will not create sound knowledge and will not be a good society.

The argument

So, the book's argument now ranges far and wide and we will sum it up again at the end of Chapter 12, but it might be useful for readers to anticipate where the book is going, so we will present a shorter version of that summary now.

Every argument rests on assumptions and our most basic assumption is, other things being equal, truth is better than lies. If you, dear reader, do not think that truth is better than lies then this book will have no persuasive appeal. The same applies if you are self-consciously committed to right-wing or left-wing dictatorships rather than what we will define as 'pluralist democracy'; the whole argument is driven by the authors' preference to see that kind of democracy understood and preserved in 'The West', if nowhere else.

We will take science as an iconic institution that aspires to make truth; whether or under what circumstances it can achieve it is another matter. We show that face-to-face communication is central to the aspiration of truth creation in science, and we argue that face-to-face communication is a necessary if not sufficient condition for all 'difficult and dangerous' truth creation, including political policymaking, and that science should offer an object lesson in respect of all these kinds of decision-making in addition to being a check and balance on political power.[5]

The internet has shifted the balance of communication in society from face to face to remote, and this brings danger in respect of what we can rely on in society and makes it vulnerable to, among other things, hidden outside influences. We show that remote communication is especially dangerous because it carries with it an 'illusion of intimacy': it can disguise itself as trustworthy local communication.

Science can give leadership to democracy because of its aspirations, but democracy has to be properly understood. Science and democracy are antithetical where the central feature of democracy is thought to be uninfluenced citizen choice (negative freedom), a model we will call 'popular assertive democracy'. This is because science is opaque to the citizenry. But popular assertive democracy rests on a false model of humankind and is self-defeating because humans are inevitably formed through their early social experience and the various kinds of continuing social engagement that follow this. 'Structured choice democracy' recognises this, resists the calls for unconstrained negative freedom and depends on elite institutions as checks and balances. The citizenry must still make the ultimate judgement at general elections in respect of whether the government and its elite institutions have done their job, but citizens must do more if democracy is to be served, or even saved; these relationships must become widely understood.[6]

----

[5] We have in mind science as an institution directed at discovering the truth of the matter. It goes without saying that there are scientists whose personal ambitions are different, that science is subject to external pressures that may subvert its central goal, and that the institution of science as a truth-seeking enterprise may be in peril as a result of the very pressures we discuss in this book. Indeed, Chapter 10 warns against abandoning the face to face in science because it would cease to be science. Truth as a goal and the face-to-face interaction that supports it are also under stress from the new ways of measuring and recording productivity in science, which have been described as a form of 'platform capitalism' (Mirowski 2018 – his 'forthcoming' is a snappy presentation of the argument). With these caveats in mind, we can still recognise the integrity of science as an institution because we can still recognise corrupt ways of conducting science as corrupt: it is when the corrupt becomes so normal that it is no longer recognisable that the institution can no longer act as an object lesson. Thus, making money out of sub-prime mortgages or high-frequency stock dealing cannot be said not to be banking because making money is the goal, not 'making money with integrity' or 'making money in productive ways', whereas science in the absence of a quest for the truth of matter is not science.

[6] Popular assertive democracy has an affinity with what is often referred to as 'neoliberalism', though that term has many meanings (see Mirowski (2019)

The illusion of intimacy and the
Law of Conservation of Democracy

There are, no doubt, many confounding political and economic factors that have led to the various political outcomes that we have witnessed over the last few years, but some of them seem inexplicable without an element of what we describe as the 'liquification of knowledge'.[7] One of the explanations of the immense power of social media to distort and 'liquefy' knowledge is a result of its having many of the characteristics of local communication yet without authentically justifying the trust we invest in local communication; we are calling this the 'illusion of intimacy'. What we call 'The Law of Conservation of Democracy' holds that populations have to understand democracy if they are to hold their leaders to account at election time for undemocratic actions.[8] If this is true then it suggests that some voters have been more influenced by slogans and digital campaigns than an understanding of, or a respect for, democracy. We will look at this in more detail in subsequent chapters, paying particular

---

for an extensive explanation of the place of neoliberal ideas in the major political movements of our time, which is compatible with the simpler presentation in terms of 'popular assertive democracy' and 'structured choice democracy' found in the later chapters of this book). For a short critique of neoliberalism and its self-destructiveness which is compatible with both Mirowski and this book's view, see Monbiot (2016). Both Mirowski and Monbiot are concerned that there is no coherent opposition position to neoliberalism such as might have been mounted by 'the left'. We wonder if such a position could be developed out of the sociology of knowledge and the fractal model of society presented in this book. This, as opposed to the incorrect and self-destructive individual freedom-based, neoliberal model, describes what humans believe as coming largely from the groups within which they are socialized.

[7] The metaphor of liquidification or liquefaction is famously associated with social theorist Zygmunt Bauman (e.g. Bauman 2000). There is some overlap between our use of the metaphor to describe the way in which previously solid claims to knowledge become more fluid and Bauman's characterization of a 'second' of modernity. There are, however, also some differences. Most significantly, Bauman uses the metaphor to refer to a pervasive uncertainty and instability that characterises of all aspects of modernity, from individual identity to globalization. In contrast, our use is more restricted and, whilst we do see the public sphere as an important site of intervention, our argument is grounded in the epistemic qualities of face-to-face interaction rather than the economic consequences of neoliberal deregulation. Gane (2001) provides a useful summary of Bauman's work.

[8] The 'Law' and the relationship between pluralist democracy and populism is explained first in Collins et al. (2019).

attention to the way the effect of digital campaigns can be enhanced by making it appear that remote communication is really local.

That a proportion of the UK population does not understand democracy very well is instanced by the success of the Conservative Party in the 2019 election in spite of Boris Johnson and his cohorts exhibiting disdain for the institutional checks and balances which are central to pluralist democracies, refusing to accept the almost-uniform expert view that a 'no-deal Brexit', which he was about to execute until stymied by a Parliament which he had tried to shut down illegally, would be an economic disaster.[9] He tried to circumvent the legal system; he is explicit about his proposed threat to the power of public service broadcasting by cutting the licence fee; and his main advisor, Dominic Cummings, is an expert at manipulating elections through the use of the social media.[10] Nevertheless, he won a landslide electoral victory in December 2019. The story of rising authoritarianism and the assault on democratic institutions and governance in the US is too well known to need repeating.

Though 'The West' celebrated victory in the Cold War with the fall of the Berlin Wall, it is not impossible that the battle is still in the balance: communism may have lost but it is not yet clear, given the new weapons in the hands of the powerful, in the West as well as the East, that it is democracy that has won; it seems increasingly likely that democracy might lose too.[11] Indeed, in some

---

[9] Which is still quite likely at the time of writing.

[10] Cummings left or was sacked from his position after the November American election.

[11] We refer throughout to 'pluralist democracy' or 'structured choice democracy', phrases which to some extent crosscut standard discussions of democracy. In political science and political sociology, the main traditions (none of which are homogenous) are sometimes described as electoral democracy, liberal democracy, participatory democracy, deliberative democracy, egalitarian democracy, majoritarian democracy and consensual democracy (Coppedge et al. 2020). We do not attach ourselves to any one of these conceptions, in part because the ideals of what we are calling democracy are found in more than one tradition. However, like Coppedge, our idealized model of democracy rests on a foundation of electoral democracy. The additional dimensions of democracy introduced here and other strands of democratic theory (e.g. deliberative democracy) depend on free, fair and reasonably frequent multi-party elections with broad suffrage. This foundation of electoral democracy is of course very old but is perhaps most famously associated with the political scientist Robert Dahl. Dahl (1971) introduced *electoral democracy/polyarchy* to describe actually existing democracies, in part because he considered true 'democracy' to be an unachievable ideal. We will return to the meaning of democracy in Chapter 11 where we will explain the distinction between popular assertive

respects it seems increasingly possible that the American Civil War is still in the balance!

The liquification of scientific knowledge

As mentioned, the Covid pandemic is a natural experiment in respect of many of the themes of this book. Among other things, it has caused the role of science to become much more salient in society and it has caused a mass shift towards various forms of remote communication. Extraordinarily, these two changes are in tension.

On the one hand the new salience of science may protect it against populist tendencies, since there is now a widespread sentiment in most countries that epidemiological science is needed to understand the spread and control of the virus, while medical science is needed to create tests for disease and tests for the presence of antibodies. The invention and mass production of vaccines is also wanted by the majority of the populations in many countries, across Europe and North America, in spite of populist interest in devaluing scientific expertise and the presence of organised anti-vaccine groups. On the other hand, the importance of science and its products is under threat from remote communication.

An author of this book (Innes) reports that in a survey of 3,696 citizens across France, Germany, Italy, Spain and UK (approximately 700 per country), conducted between 18 March and 30 April 2020 (i.e. when the Coronavirus pandemic was at its height), over half of those questioned reported having encountered disinformation and/or fake news about Covid-19 online. Intriguingly, across all five countries, those who had seen such material were more likely to believe that disinformation impacts trust in science, experts and health policy 'to a great extent', and they also believed it impacts upon trust in government.

In a UK survey (https://www.kcl.ac.uk/news/whos-most-likely-to-refuse-a-covid-19-vaccine), 27% of those who say they get a great deal of information on Covid-19 from WhatsApp say they are unlikely to or definitely will not get a vaccine, and younger people are more hesitant or resistant, with 16–34s (22%) twice as likely as 55–75s (11%) to say they are unlikely to or definitely will not get a vaccination if it becomes available.

Reported results from the US indicate that around 34% of the population might not choose to undergo vaccination even when it is readily available, though the percentage varies with age (the young being less inclined to vaccinate), with politics (Republicans being less likely to vaccinate) and with location, (those from the South being less likely to vaccinate).[12]

---

and structured choice democracy, with pluralist democracy being a subset of the structured choice variety.

[12] Source: Ramjug (2020). There are, no doubt, structural causes of the increasing liquification of knowledge, such as the shift to the right of 'Western'

It might seem that all is well so long as the large majority in these countries would still choose vaccination, but it must be borne in mind that vaccination works better the more people choose it because this reduces potential sources of infection for all (or even eradication through herd immunity), and the numbers reported are not reassuring in this regard.

These results in respect of Covid reinforce the already-developed argument in the pre-pandemic draft of the book in respect of the resistance to the measles, mumps and rubella (MMR) vaccine. Among the authors of this book, Collins and Evans have argued that the MMR vaccine-rebels, especially parents who, encouraged by the mass media and social media, effectively took themselves to be experts in the matter in consequence of observing the autistic trajectory of their own children, and who took notice of the celebrities, doctors and politicians who championed and publicised the anti-vaccine views, were seriously misled.[13] Now we know as a result of work by Innes and his colleagues that, more recently, Russian disinformation and misinformation techniques have been applied to coronavirus (COVID-19) and were applied to the MMR case and that these, along with the general reinforcement from social media networks, were probably responsible for at least some element of the continuation of the MMR revolt and the considerable number of measles cases that the loss of herd immunity has created. We will return to both the MMR and the Covid case later in the book, using MMR as a 'hard case' when we get to the final chapter, which asks the difficult question about what to do next.[14]

---

politics, perhaps engendered by the failure of Western democracies to maintain something like the economic 'Post-War consensus' and instead foster increasing economic inequality. For example, Kennedy (2019) argues that support for populist parties and vaccine hesitancy in European countries are both driven by a distrust in experts and elites that can be linked to economic and political marginalisation. This is possibly reinforced by the post-modernist movement, that works in interaction with the cause we identify here: the increased use of social media and the displacement of face-to-face communication with remote communication. But our subject remains communication and liquification.

[13] Unfortunately, some STS scholars, who believe that one of the consequences of the liquefying of knowledge is that science should be 'democratised', championed the rebel parents. Collins and Evans (2017b) argue that the MMR vaccine is a vital case study for how knowledge is going wrong. Brian Wynne, Melissa Leach and Brian Martin are social scientists who either argued directly or mounted arguments in respect of other medical controversies that aligned with the view that the parents were right. This material is diffuse but we can report several confrontations at conferences, workshops and seminars, and see https://doi.org/10.1016/j.socscimed.2004.12.014 for a sense of the 'logic' of the defence of the 'underdog' parents.

[14] Examples of Covid disinformation will be found in Appendix 2.

But the other side of the COVID case – which does not reflect the MMR case because MMR did not cause a lockdown – will also be revisited. The Covid lockdown has given rise to what seems to be something like a 'social movement' within science to close down the very face-to-face meetings that we are arguing are essential if the formation of scientific knowledge, and science as an institution more generally, are to remain distinct from the kinds of opinion formation that is common in social media – in other words, if the basis of scientific expertise is to remain something other than the number of 'likes'. These matters will be revisited at length as the book unfolds.

Trust and the face to face

Trust and communication are like conjoined twins: to know how to act on a communication – how much to risk – you have to know how reliable it is. We will argue that trust and communication are, in turn, conjoined, with a third sibling – face-to-face interaction – and, through this, to the small group. It is obvious that face-to-face communication is most natural in small groups, and we have also to take into account that we tend to trust those we know, and those who are like us – which is often the same thing. This is known as 'homophily'. On the deep origins of this tendency we are not experts, but we can see that solidarity among families and solidarity among tribespersons, with children being continually warned not to trust strangers, would be a natural starting point. We might even cite the size and capacity of the human brain, which, seemingly, limits our deep relationships to small numbers, 150 sometimes being said to be a rough upper limit.[15]

Science, which we do know quite a lot about, works, as we will explain, with small groups, and science's aspiration is above all, to try to make reliable knowledge – to find the truth. So, there is a set of interconnected arguments that, when we want to understand the origins of sound social interaction, take us back to (a) small groups, whose members (b) trust each other, (c) communicate with each other and (d) communicate directly and in person – these are conjoined quadruplets. In science we know that such groups are influential in forming what they believe will come to count as truth.[16] We also know that such small closed groups can also work for ill, for example, maintaining the boundaries of elites where they are power-seeking rather than truth-seeking,

---

[15] For the theory of the upper size of human groups as limited by the capacity of the brain seen Dunbar (2012). (Smith et al. (2020) discuss this work in the context of a larger body of research.)

[16] This idea is not identical to but is close to what are known as scientific 'core-sets' – a core-set could include competing groups who do not trust each other. Steven Shapin (1994) was one of the first to stress that science depends on trust.

persuading cult-members to follow a leader's dangerous path, or providing the tight boundaries and trust needed for crime or terrorism, but the argument is asymmetrical: as stated above, small groups and face-to-face communication in small groups may not be a *sufficient* condition for creating the best kind of knowledge but they are a *necessary* condition. We are going to argue that where we look for the best kind of knowledge, it is the small group that will form the basis for it even though, in other circumstances, it could form the basis for less-desirable kinds of knowledge. Whenever we say 'this depends on face-to-face communication in small groups', that is what we mean; we will not repeat the 'necessary not sufficient' *caveat* every time we say it because it would become tedious.

The sociology of knowledge

At the risk of over-simplification, the reason there is even a question about what knowledge we trust – the reason that the truth is not self-evident – could be said to be the problem of induction and its sceptical philosophical penumbra. How do we know the world we experience will continue beyond the next moment; what is to stop everything being different next time we look up? In sum, how do we know anything enduring? How can we rely on anything more than our momentary sense impressions and how do we put these momentary sense impressions together into a world of more substantial objects and processes? The *sociology of knowledge* points out that different societies put things together in different ways. Momentary sense impressions aside, everything that we think we know in a secure way, we know from unreflecting acceptance of what we absorb in the course of our upbringing and, as adult life unfolds, our continuing socialization. Even the things we count as secure scientific knowledge we know from what we have been told, unless we are among that tiny number of people who have done the relevant experiments or observations and even those experiments and observations depend on trusting all the others who have made and interpreted what goes into them.

In the traditional account of the Enlightenment, religion and royalty offered a solution to the problem by providing an unquestioned source of authority upon which societies could be built, shared consensuses could be established, and moments of uncertainty could be stabilized. Science fought to unfix these things and make at least some knowledge open to change through theorization and experimentation by people who were neither divine nor royal; science made knowledge a bit more malleable. To introduce a metaphor that will be useful throughout the book, science made knowledge a bit more 'liquid' than it was in more traditional societies. But scientific knowledge was still taken to be pretty solid – scientists became new sources of authority, able to claim that scientific method stood above society in the place that religion and royalty had once occupied, and scientific knowledge was nearly as fixed; only on rare occasions did scientific knowledge change in significant ways.

Knowledge becomes liquid

In the second half of the twentieth century the *sociology of scientific knowledge* was invented; even scientific knowledge was shown to be a lot more fluid than had been thought.[17] The sociology of scientific knowledge, actively or unwittingly, contributed to the larger movement called 'post-modernism', under which, in theory at least, every kind of knowledge became fluid.

In the early days, the sociology of knowledge (of scientific knowledge and of knowledge in general), and the still more ambitious post-modernism, were esoteric concerns bounded by the world of academe, and they had little impact outside that world. Different academics from the humanities and social sciences located themselves somewhere along the spectrum of academic possibility according to what we might jokingly call their 'knowledge liquidity preference,' and argued with each other about where you had to be on the spectrum in order to understand knowledge properly. For some, facts were still objective, neutral and dependable, revealing real things about the world around us. For other scholars, the priority was demonstrating ways in which what seemed to be firm knowledge was merely frozen liquid: for sociologists the important thing was the role of social interactions in constructing which piece of knowledge was believed and which was not, and how values and biases were introduced by researchers.

The sociology of knowledge is dizzying because it continually digs the ground from under one's cognitive feet. It is hard to remember that our knowledge of even the commonplace, like why aeroplanes fly and why wine intoxicates, are the result of our upbringing and trust in what we hear said in day-to-day life; the source of our certainty is the same as that of our one-time certainty about witches and devils.[18] Our knowledge about aeroplanes and intoxication does not arise directly out of what is still the most promising source of sound and enduring knowledge, flagging under various assaults though it may be: scientific method; with only very rare exceptions, people who feel they know these things about alcohol and flight have generally done no empirical examinations of aerodynamics or fermentation and physiology. Still, the almost irresistible belief that we have a better kind of certainty than we used to have about witches and devils is comforting, in a way: it shows that the traditional world of trust in

---

[17] Collins, one of the founders of the sociology of scientific knowledge, invoked the metaphor of liquid in his 1985/1992 book, arguing that scientific knowledge should be seen as ice-cream. The ice-cream of scientific knowledge is normally kept frozen solid by the activity and affirmation of scientists inhabiting their scientific social groups, whereas left to itself it will lose its form. 'Heat or pressure – representing the revolutionary or extraordinary periods of science – will turn it rapidly to liquid.' (p. 184 of 1992 edition)

[18] The examples are taken from heated arguments between Collins and university-based philosophers who claimed their certainty in these things was based on something more direct.

the respected institutions of knowledge – in this case trust in *other people's* use of scientific method – is still with us to some extent, even if knowledge is not quite so stiflingly fixed as it was in truly traditional societies. *On our days off from academic analysis,* we still live in a world in which we can usually rely on the fact that aeroplanes fly and wine intoxicates, and both will continue to do so for what are generally counted as good reasons; it is still, in many respects, a traditional world. And there is no other way it can be – because no-one can investigate very much of what they rely on directly: to have a stable world we must have a traditional kind of trust in certain institutions.

But these traditions are fading as the once esoteric preoccupations of sociology of knowledge and post-modernism become everyone's day-to-day business. The reasons for believing in the liquid end of the knowledge liquidity spectrum was always intuitively grasped by propagandists, by George Orwell and his like, by holocaust deniers and by other purveyors of 'the big lie' with Nazi Germany as the icon. This understanding has found its way to the powerful once more.[19] What is new, as we explore in Part II, is the way remote forms of communication such as the internet and social media have amplified and accelerated the everyday liquification of knowledge, taking the process out of the academy and putting it into the smartphone.[20] The liquidity of knowledge has become everyone's day-to-day business. Nationally organised political disinformation techniques, which we will discuss in Part II with some Russian examples, may not seem to affect the ordinary citizen's day-to-day life, but insofar as they bear on the argument over the safety of vaccines – which, as we will see, they do – they will, almost certainly, affect the health, or even survival, of every individual or someone close to them before long.[21]

Dizziness

But here is a problem. All of the authors of this book abhor and decry the re-emergence of measles epidemics as a result of the loss of herd immunity

---

[19] There is an ongoing debate in science and technology studies (STS) circles about the extent to which this kind of academic work is responsible for the understanding of the liquidity of knowledge now being exploited by the powerful to the detriment of environmental concerns or tobacco users (eg. Oreskes and Conway, 2010). Many in STS circles are engaged in a scramble to excuse themselves from any responsibility (Collins, Evans, and Weinel 2017; Jasanoff and Simmet 2017; Lynch 2017; Sismondo 2017a, 2017b).

[20] For a rich and chilling account of these developments see Pomerantsev (2019a).

[21] A sign of the times is that the scientists who discovered the first gravitational wave had to take into account that the putative discovery might have been the result of the instruments being maliciously 'hacked' (see Collins (2017) for an account).

to measles due to the non-take-up of MMR. But now we, the authors, have to ask ourselves a difficult question – the same question that we complained about when we explained how hard it is to cleave to a sociology-of-knowledge view of the world: 'Given the liquefaction of knowledge, how – and how robustly – do we know what we know about the MMR case or the recent epidemics of measles?' We have to ask ourselves this with even greater urgency because we now know the power of social media and the internet – we know that society's knowledge in general is becoming more and more liquid. And that applies even to the things we think/thought we knew through our privileged understanding as academics. How do we know that the pro-vaccine case does not itself turn on a conspiracy just as we claim that the anti-vaccine case turns on a conspiracy? None of the authors are epidemiologists who have examined even a single measles outbreak in person, and we do not even know any children with measles![22]

Unfortunately, the answer to how *we* know what we know, and how anyone else should know what they know, is not an easy one. But if we cannot find an answer then scientific expertise will eventually disappear and societies as we know them will fade away; when the very notion of experts and expertise disappears, there will be nothing left to guide the choices of even the most ideal democratic governments except riches and power and acclaim.[23] This book is, therefore, trying to make a contribution to the answer to the question of who to trust. It will not be a sharp answer like the solution to an equation; it will be an attempt to explain something far less precise about the changing nature of knowledge and something that will put us in a position to think more accurately about what we can do about it. The answer is going to turn on which institutions we trust in our society – a matter of citizens' meta-sociology – and that answer will be justified by how those institutions aspire to make their knowledge.

---

[22] Many of these points about MMR and measles could be made with the much more immediately relevant case of coronavirus (COVID-19). The trouble is that uncertainty about Covid-19 is more immediately a matter of the uncertainty of the science, and this confounds the questions we want to ask in this introductory part of the book, whereas there is a very strong consensus these days about MMR and therefore this helps the focus on sociology of knowledge questions. We are going to stick to MMR as our *hard case* in terms of the questions we want to ask here.

[23] Collins et al. (2019) explain that pluralist democracy differs from populism in its dependence on checks and balances and these depend on the acceptance of domains of expertise – most obviously in the case of an independent legal system. Scientific expertise is also one of the checks and balances. We will return to this in the final chapters.

What kind of society do we prefer?

As this portion of the text was being written, President Donald Trump was denying the charges underlying potential impeachment concerning his pressuring a foreign government to investigate his political opponents; in the UK the government was appealing directly to the electorate so as to nullify the authority of Parliament. All this is taking place in the context of 'alternative facts' and 'post-truth'. Even in 'Western' countries, who to trust and how to make up one's mind can no longer be seen as something that past generations had fought wars to resolve once and for all; the dilemmas have re-emerged and become our problems too. The question is whether we want to live in a world where decisions on healthcare, social policy and international affairs are informed by those who have studied them in depth using as much scientific expertise as they can muster, and where politicians are selected for their capacity to advance policy informed by the best available knowledge, or whether we want to live in a world where knowledge is created moment by moment in the spin of the internet liquidiser.

Measles, Mumps and Rubella (MMR) vaccine as a hard case

In the kind of society we favour, citizens would not have been taken in by the scientifically flawed and/or fraudulent case against the measles, mumps and rubella vaccine (MMR), and would recognise the epidemiological argument for MMR's safety and necessity when set beside the first-person reports from parents whose children seemed to exhibit symptoms of autism shortly after vaccination; citizens would accept that this second kind of 'evidence' was not convincing and they would not be swayed by celebrities and politicians; they will seek Covid vaccinations once the proper tests have been completed.[24] But, as we will see, the MMR case is a really hard one because of the way it involves the public and the way the scientific expertise was published and discussed by the press, which failed to act as a check and balance on scientific claims but preferred the journalistic norm of 'balancing the story'. This, ironically, is why MMR is so valuable as a case study: it is a 'hard case'. It is not a trivial matter to work out how to accomplish a society that works properly in a case like this. It will turn on changing what is 'taken for granted' in society as a whole.

---

[24] Unfortunately, some humanists and social scientists are primarily concerned with championing the postmodernist liquidity of knowledge as an end rather than an analytic technique. At the time of writing, Putin's Russia is claiming to have developed a coronavirus (COVID-19) vaccine, but it has not been tested to the standards of 'The West', in which tests take a long time to complete. (At the time of publication, the Russian vaccine appears to have passed the tests typical of Western vaccines.)

Among the things we are trying to do here is provide yet another kind of argument for why the ordinary person should, in the first instance, trust the institutions that represent scientific and non-anecdotal, experience-based expertise, when faced by competing accounts.[25] If we want the kind of truth-based society for which we have expressed a preference, we will all have to come to understand that, though science and technology and all other kinds of technical expertise, are always flawed and fallible, abandoning them is a certain route to dystopia. The MMR vaccination example is at one end of a thesis about the possibility of a society based on the idea of the truth. A society is defined by what people count as true without thinking about it. We argue from how early on, socialization works to make a society – it creates the 'taken-for-granted' – to a concern about how the taken-for-granted is being assaulted by social media, surveillance capitalism and the micro-targeting of messages. Once, tradition was the hardest thing to change and this, of course, was suffocating, but now it is too easy to change, and we have no firm jumping-off point for forward movement.[26] We aim to explore how undue emphasis on the *remote* can accelerate a transition away from the kind of society we want to inhabit and why, therefore, we must understand the importance of the face to face.

---

[25] The justification of science is a project with a long history going back to science's foundation. In relatively recent times – the early part of the twentieth century – it was thought that science was so obviously right that the only thing to do was explain why. But the philosophers disagreed, for example, some thinking that experimental verification was the key and others thinking, by the 1950s, that the important thing was vigorous but failed attempts to falsify a claim. Historians and sociologists, examining science very closely in the 1960s and 70s, showed that the results of any kind of experiment could be interpreted in various conflicting ways, so the outcome of such a passage of activity was much closer to political preference than had been believed. In response to this, two of the current authors (Collins and Evans 2002, 2007) argued that those with expertise should be trusted but that it has to be expertise informed by certain values; luckily those values are consistent with the overarching nature of scientific activity – the search for truth (Collins and Evans 2017b). Part III of this book has turned out to be, among other things, another step along the same road, with it being argued that new truth is best sought in face-to-face interaction in small groups and, since scientists work that way, that is another reason they should be trusted.

[26] The diminishing importance of traditional institutions and life-course pathways, and the increasingly individualised world that follows from this, is a key feature of the risk society literature (Beck 1992; Giddens 1990). It is particularly strong in the work of Anthony Giddens, where the term 'post-traditional' is used to capture the ways in which individuals must reflexively craft and manage their identity through choices made under conditions of ever-changing information (Giddens 1994).

The chapters and main arguments

The book, as explained, is divided into three parts. Part I is called 'Foundations: Communication, Socialization and Trust' and has three chapters. It explores and tabulates the importance of face-to-face interaction from the constituting of society and the individuals within it at the most profound end of things, to hard-to-reproduce efficiency at the other, all listed in a table that is central to the book – Table 1.1. Chapter 1 pulls apart the idea of 'trust' into four different types introducing an important distinction between 'trust' and 'reliance', terms which are often used interchangeably; reliance, as we use the term, is the taken-for-granted version of trust, and the distinction is vital for understanding what happens in different forms of communication. 'Trust', unavoidably, takes the role of the generic term covering all the meanings including reliance, but this should not cause problems. Chapter 1 goes on to introduce the 'fractal structure' of society (represented in Figure 1.2); it is based on the continuing contribution of socialization from infancy to the acquisition of expertises throughout adulthood. Table 1.1 and Figure 1.2 will reappear throughout the book and at the end as the basis for Table 10.1 and Figure 12.2 respectively. In Chapter 2, we explore the foundational importance of trust and why infant socialization offers a first insight into the importance of face-to-face socialization. By the end of Chapter 2 we have worked half-way through Table 1.1. In Chapter 3, 'What Face-to-Face Communication Offers', we complete the exposition of Table 1.1 by discussing the features of face-to-face communication that are to do with efficiency and the lack of brittleness of direct communication. By the end of this first section, the reader should have a clear understanding of what face-to-face communication is, why it is central to trust and trust is central to it, and why it also has less profound but very difficult-to-replace advantages in terms of quick and safe transmission of meaning.

Part II is entitled 'Arguments and Evidence: Can Remote Communication Replace Face to Face?' Part II explores the weaknesses of the remote, an integral part of which is exploring its strengths and explaining why they do not add up to a general solution for the future of communication. Part II, then, takes on the impossible task of trying to prove a negative – that remote communication can never replace face-to-face communication. The task is impossible because to prove it decisively requires prophecy not science: it would require us to know the future of all technological developments that bear upon communication. But the perfect cannot be allowed to drive out the good and the impossibility of a final proof of a negative cannot be allowed to rule out critical analysis; what we can do is consider the foreseeable future using as much imagination as we can muster. The nature of the task means that the arguments and evidence reach in many directions and that is why there are six heterogeneous chapters in Part II and why the middle part of the cognitive funnel is an elongated and irregular cone. Additionally, the five appendixes mostly relate to the material and arguments found in Part II but we have separated them out for the sake of readability. They are about propaganda, a supplement on disinformation, citizens' understanding of science, cases where remote communication does work in science and technology, and second language learning. It would have been possible to fit them in to this section but presenting them as appendixes enables readers to choose more easily whether they need to read any of them while making clear that we have not overlooked certain potential counter-arguments.

Table 4.1 tabulates types of remote communication though it cannot be as straightforward as Table 1.1 because remote communication technologies can be used in many different ways. Chapter 4 begins by introducing this table before exploring the circumstances under which remote communication can be especially productive. Chapter 5 explores, more speculatively, the future of remote technology asking whether technological progress will eventually supersede these critiques. Chapter 6 explores how it is that, since face-to-face communication in small groups is claimed to be so central to trustworthy communication, we can possibly have mass societies that depend on remote trust and whether their very existence defeats our central argument. We look at the various explanations for how large-scale societies grow out of small groups and how long-distance trust is maintained in modern societies, and conclude that none of this is incompatible with what we have argued when it comes to the 'difficult and dangerous trust' that underlies new and crucial knowledge. Chapter 7 presents a series of case studies of communication in circumstances that one might think would be ripe for the transition from face-to-face to remote communication and shows how 'sticky' the face to face is even where there are obvious advantages to change. Chapter 8 explores why it is that remote communication can be misleading and dangerous; in part this is because it can appear to be communication originating in a local group even when it is comes from far-right recruitment groups or even coordinated disinformation and misinformation campaigns managed by foreign states. Here we develop the idea of the illusion of intimacy in more detail.[27] Chapter 9 looks at the use of the internet in deliberately misleading disinformation campaigns.

---

[27] Turkle (e.g. 2011, 2015) has long studied human-machine relationships as a psychoanalyst, basing her research in the MIT Artificial Intelligence Laboratory. One of her books describes her project as dealing with the 'illusion of companionship', which is generated, notably in children, by interacting with robots that maintain eye-contact and with other machines that try to interact conversationally. She also looks at the dense 'conversational' interactions that take place with social media and which distract us from face-to-face interactions. She claims that the environment created by these new patterns of interaction with machines can distort children's and adults' understanding of real emotional relationships and erode their ability to engage in them. Our approach is less individualistic and psychological: we take knowledge-making to be a collective activity and we are concerned with how remote communication effects this collective ability rather than

Part III is called 'Conclusions: Science, Truth, Democracy and the Nature of Society,' and sets out the dangerous consequences of the shift from local to remote communication and what might be done to alleviate them. Chapter 10 deals with what we see as the misplaced demand, inspired by the pandemic lockdown, for face-to-face scientific conferences to be abandoned in favour of remote collaboration. It explains that this would be disastrous, given both the place of science in our argument and the central role of face-to-face communication in science. An analysis of the way scientific conferences work, instead, leads into suggestions for how the conference circuit could be changed to reduce its negative consequences but without abandoning the opportunities for the face-to-face interactions that are crucial if science is to continue to work towards the creation of truth and fulfil its vital role in any democratic society. Chapter 11 widens out further to analyse the nature of democracy, among other things introducing a simple division between 'popular assertive democracy' and 'structured choice democracy' (see Figure 11.1), which cross-cuts much traditional debate about democracy's nature and sets up the claim in Chapter 12 that popular assertive democracy is unsustainable. Chapter 12 asks what is to be done? It uses the measles, mumps and rubella vaccine as a hard case for hammering out possible answers and relates these to the current state of the political world. Finally, there is a Postscript discussing the significance of the 3 Nov 2020 US election result.

The way remote and social media technologies on the one hand, and face-to-face communication on the other, are used and the way they relate to each other, has the potential to change the way we make decisions and choices. Where Part II established that remote technologies have shortcomings when they try to replace the entirety of face-to-face interactions, Part III argues that remote technologies can actually be even more pernicious in the creeping ways they reorder democratic societies. We suggest what needs to change if bad effects are to be avoided.

Taken together, we offer a case not for eliminating remote communication, something that would be impossible and crazy given its benefits, but for (a) being aware about where face-to-face and remote methods can each be most advantageous, and for (b) being cautious and prepared for the negative ways that remote communication can affect the very constitution of our society and the things we think we know. These are critically important issues, and we must not allow the convenient to overpower the essential.

---

how it affects emotional well-being. The difference can perhaps be made still more clear if we consider that many of things that Turkle argues, such as the fact that one cannot *earn* the love of a machine or a digital 'friend' in the way one can earn the love of a human, might equally apply to our relationships with animals (e.g. there is no risk of betrayal in the love given by a dog to a human), whereas animals play no part in the construction of trust- and language-based knowledge.

PART I

Foundations: Communication, Socialization and Trust

CHAPTER 1

What Trust and Communication Are For

The uses of face-to-face communication

We now begin the analysis of the face to face, the results of which are anticipated in Table 1.1. This table will be a constant resource throughout the book and is, in many ways, its core. We are arguing that science and scientific expertise are key institutions for the survival of democracy, that face-to-face communication is a necessary feature of science and, therefore, that trust in knowledge that is grounded in face-to-face communication is a vital feature of democracy. Figure 12.2, which we have described as a summary of the book's overall argument, is a graphic presentation of what we think happens as the face-to-face communication that supports scientific and other expertises is replaced by the remote communication of social media and the like. A key part of our exposition, then, must be to explain what makes face-to-face interaction special, and Table 1.1 is a summary presentation of these qualities. We will also use Table 1.1 in Chapter 10's Table 10.1, to develop a classification of face-to-face interactions.

Table 1.1 will be explained over the next three chapters, but it seems appropriate to present it here, at the beginning of the book. Part II of the book will introduce a table describing kinds of remote communication though it will be a little different in concept to Table 1.1. This chapter and the next will take us halfway through Table 1.1 while Chapter 3 will complete the exposition.

Table 1.1 lists twelve features of face-to-face communication in four groups. The divisions into twelve features and four groups are to some extent arbitrary as the borderlines are not always sharp, but nothing of significance would be changed by dividing things up a little differently. The four groups are listed in rough order of descending profundity, and it is not surprising that two chapters will be needed for the first two groups and first six features, while the job will be completed with just one additional chapter.

How to cite this book chapter:
Collins, H., et al. 2022. *The Face-to-Face Principle: Science, Trust, Democracy and the Internet.* Pp. 23–43. Cardiff: Cardiff University Press. DOI: https://doi.org/10.18573 /book7.b. Licence: CC-BY-NC-ND 4.0

Table 1.1: Twelve features of face-to-face communication in four groups.

GROUP		FEATURE
I	Forming society	1. Expressed commitment and injection of energy 2. Learning from the bath of words 3. Acquiring reliance (cf trust) 4. Tacit knowledge transfer
II	Trusting individuals	5. Learning how to trust (cf reliance) 6. Domain discrimination
III	The use of presence to create and modify meaning	7. Immediate influence on interpretation 8. Body language modifying meaning 9. Safer adversarial dialogue
IV	The presence of the body to promote efficiency	10. Efficiency in conversational turn-taking 11. Efficiency in number of meetings in one location 12. Serendipitous meetings

The first group in Table 1.1 involves the formation of the collective basis of human societies. Since we have newly forming societies in mind, the first item is the sheer energy and commitment that is generated by gathering people together in one place, especially if they come from a distance.[28] But in due course we will argue that the legacy of growing up together in one place – the family and the tribe – may be the basis of the solidarity that is associated with co-location. The first group also includes learning from the bath of words, acquiring *reliance*, and sharing and building tacit knowledge. The meanings of these terms will be explained shortly.

The second group is about *trust*, as opposed to reliance, and includes learning how to trust people and not trust people in general and learning how to do this in bounded technical domains (domain discrimination). The third and fourth groups build from the first two. In some ways, their content is less deep, but their impact is important and more immediate. The third group is to do with the use of physical presence and body language to help transmit and to modify meaning in local groups, particularly to use the body and the mode of expression to disagree firmly but safely. The fourth group is about the various efficiencies associated with co-location including the subtle management of conversational turn-taking, and the co-presence of people making for easily arranged and fortuitous meetings.

---

[28] Our colleague Alun Preece points out that though the commitment is much less than that involved in travel to a remote site, some commitment is necessary even to spend an hour or two on remote conferencing and this is valued – so remote communication is not entirely devoid of the expression of commitment.

Ways of trusting

We now start on the detailed exposition. The world is based on trust: without trust we would not dare to leave our front door and we would not be able to believe a word that was said to us; as already mentioned, communication is conjoined with trust. In this section of the book, we are going to divide trust into different kinds with different labels, but the term 'trust' is so embedded in the English language that, from time to time, we will find ourselves using it as a generic term encompassing all four meanings including reliance: this very section heading is an example of the problem. Readers should, however, have no problem recognizing what is intended.

The Cambridge English Dictionary defines 'to trust' as 'to believe that someone is good and honest and will not harm you, or that something is safe and reliable'. *The Cambridge Dictionary* also treats the term 'reliance' as a synonym for trust, but we will define reliance as something different. In the meantime, look at the *Dictionary*'s common definition and note that there are two distinct meanings present already: 'to believe that someone is good and honest and will not harm you' involves a judgement to do with the moral world and others' internal states; to believe that 'something is safe and reliable', something like a bridge or an aeroplane, is a judgement to do with the material world.[29]

One can find this distinction in the existing literature, sometimes signified by the terms 'trust' and 'reliance': trust is the moral category whereas reliance is the material category.[30] But we think there are at least four distinct and important meanings of trust. We think the trust in bridges and aeroplanes is better described as 'dependence' rather than trust, while the term 'reliance' is a better fit for a much more important category not discussed in the literature, to which we will come in a moment. The third category is another important one for this book and fits between trust and dependence as we have defined them; one can

---

[29] Sometimes trust in bridges and aeroplanes turns on trust in the people who built them so this category could be split into two, but we will not develop this point.

[30] The *Stanford Encyclopedia of Philosophy* (McLeod 2021) uses 'reliance' to mean something like our trust in the strength of bridges and the like; it uses reliance to mean the kind of trust that cannot be 'betrayed' only discovered to be misplaced. In a similar vein, Luhmann (1998) distinguishes between trust and confidence, where confidence is defined as relating to settings where you do not consider an alternative and trust is used to characterise those settings where a choice is made. The moral dimension of trust is revealed by what happens when expectations are not met:

> In the case of confidence you will react to disappointment by external attribution. In the case of trust you will have to consider an internal attribution and eventually regret your trusting choice (Luhmann 1998, p. 97–98)

believe that a person is safe and reliable not because one has made an assess-ment of their moral integrity but because one believes things have been arranged in such a way that their self-interest will keep them behaving in the way one expects. It is a kind of judgement concerning the material world, with humans treated as material objects whose behaviour is mediated, not by physical forces but by what economists call 'utility functions'; both physical forces and utility functions can be expressed in equations rather than being a matter of moral judgements. An employer, for example, by paying an employee a wage for cer-tain services – the necessary level of which, it is assumed, is widely known and accepted across all work of that type – modifies the employee's utility function in such a way that, *as they see it*, the employee can be 'trusted' to continue to ren-der those services. The employer and employee enter into *a transaction* and this is what we will call 'transactional trust'. To repeat, we say it lies between trust to do with the moral world and trust to do with the material world because, in terms of trust, it treats the human more like a material object than a moral object even though it is based on modifying humans' internal states.

Of course, there are different ways of instantiating such a relationship, and different approaches to modifying utility functions, from slave-labour encour-aged by punishment to the trusted 'family retainer', the former being at the far material end and the latter being at the moral end. We may, for example, imag-ine a family falling on hard times but the trusted family retainer continuing to serve, seeing themselves more as a member of the family than an employee. We have, then, a distinction between transactional trust and 'moral trust' which will turn out to be key to understanding some cases of remote trust.

In sum, 'trust' is the generic term covering:

(1) *moral trust* in persons
(2) *material dependence* on physical objects
(3) *reliance*, which, as we will shortly explain at greater length, contrasts with trust in that the former is taken-for-granted and unconscious while trust is thought through
(4) *transactional trust*, which treats persons' trustworthiness as more like material dependence.[31]

---

[31] Shapiro (2012) develops a set of meanings of trust that arise out of eco-nomic thinking and have to do with the likelihood of being right, including that the trusted person's interests coincide with those of the person doing the trusting. Shapiro, as we will see, is primarily concerned with 'trustees' – those who are placed in official positions of trust over others' property or well-being. A related set of distinctions can be found in Atkinson (2007, p. 228).

Transactional contracts have been defined as specific, monetisable exchanges between parties, the focus being on providing monetary remu-neration for services provided by the employee (De Meuse et al. 2001) and

Face-to-face interactions and remote interactions depend on these different kinds of trust in different ways.

Just for completeness, there is another category, which we will not add to the list because it will not appear again in the argument. This is treatment of a human as untrustworthy because, though they have integrity and their utility functions are not suspect (e.g. they are not under threat or subject to bribes and are not suspected to have 'ulterior motives'), they are considered to be insufficiently competent to carry out what they promise. This is, again, treating a human as a kind of material object with a faulty design.

The difference between trust and reliance

When self-conscious trust proves unfounded it leaves the one who has trusted feeling foolish because they have made a miscalculation or a misjudgement; they will feel stupid if they have chosen to depend on a bridge which is unsafe or a human who is unreliable; they will feel *betrayed* when the judgement was a moral one. Reliance, on the other hand, is simply part of the fabric of our lives and enters our consciousness only if there is a studied effort of reflection typical of philosophers, sociologists and the like. A baby, for instance, *relies* on its mother without understanding the relationship; if we simply say 'a baby trusts its mother' we are not capturing the special nature of the interaction because babies do not know how to *trust* or what *trust* is – they do not know how to judge others' internal states or that there are such things. Adults rely on many things in the same way that babies rely on mothers, even though adults have the potential to reflect if they have the right kind of training. For example, an adult relies, without reflection, on the future being pretty similar to the past unless they have been introduced to the philosophical problem of induction, in which case they will become more aware of what their regular actions imply in terms

---

establishing the notion of 'a fair day's work for a fair day's pay' (Rousseau and Wade-Benzoni 1994). Such contracts are premised upon economic exchange perspectives (Rousseau 1989). Relational contracts, however, are presented as open-ended, less specific agreements that establish and maintain a relationship, being based on emotional involvement as well as financial reward (Robinson et al. 1994) and emerging from social exchange perspectives (Blau 1964). These definitions present transactional and relational contracts as being separate entities, however, more recently it has been suggested that this may not be the case and that there may be a continuum along which a contract is based, rather than it being simply at one of the two extremes (Coyle-Shapiro and Kessler 2000).

(As the quoted paragraph is presented as an illustration of a kind of thinking, the bibliographical references within the paragraph will *not* be found in this book's bibliography.)

of reliance. But adults also rely on not being attacked on the street and everyone driving on the correct side of the road and shopkeepers exchanging the chosen goods for money, and so on. The baby example aside, these are instances of culturally generated trust, trust which grows without self-conscious realization or calculation as one becomes a member of this culture or that. When an adult discovers that an instance of reliance has proved to be misplaced, they do not feel foolish or betrayed, as in the case of self-conscious trust; rather, they discover that the world is not the place they thought it to be.

The distinction between trust and reliance is vital if we want to understand how societies work. One society differs from another according to what individuals within the society rely on *without* thinking; this is part of what sociologists call the 'taken for granted reality'. Do individuals in that society rely on magic? Do they rely on medical doctors? Do they rely on cab drivers?

Notice the word 'cab'. It is intended to bring to mind something like the London black cab or the New York yellow cab that can hailed from the street. Traditionally there has never been any question of trust involved here – when hailing a cab one is not thinking about whether the driver can be trusted, because the institution is *relied on* (not 'trusted').[32] Uber is a bit different because it is a different kind of institution – one that we are still coming to terms with just as, in Britain, at one time, we had to come to terms with so-called 'minicabs' and, in fact, have never come to rely on them in quite the same way as we rely on traditional cabs. In the case of Uber and minicabs, thinking *trust* has, to some extent, displaced unthinking *reliance*. One of the disturbing things that is happening to Western societies is that more and more institutions that were once simply *relied on* are turning into institutions that involve self-conscious choices about trust. For example, we no longer simply rely on the banking system but ask whether we can trust it and, insofar as we must trust it, we find ourselves estimating which banks are the most trustworthy. Fifty years ago, in Western societies, all banks were equally reliable as far as security of funds and, say, the small cost of accidental overdrafts were concerned.[33]

The sociology of trust and reliance

Table 1.2 sets out a couple of the distinctions that relate to what has just been discussed though the difference between short-term and long-term criteria will not be fully explained until we get to Chapter 4: Remote Technology and Trust.

---

[32] Of course, reliance can be violated on rare occasions, but they are usually so rare as to make headline news and the plots of books and films.

[33] One of the authors (Collins) transferred his account from one bank to another because of a huge charge on a tiny accidental overdraft lasting a couple of days and at the time of writing (June 2019) the UK is having to introduce legislation to stop this kind of cost-gouging.

Table 1.2: Trust and reliance in short-term and long-term.

Self-conscious vs. tacit / Criteria of Success	Self-conscious TRUST	Tacit RELIANCE
SHORT-TERM	1	3
LONG-TERM	2	4

In cell number 1, we find examples of short-term, explicit trust – anything where you take a self-conscious decision to trust something and will soon find out if that trust has been justified; buying a used car is as good an example as any. Where there are well defined criteria of success, we have clear cases of trans-actional trust. These will resurface in the discussion of remote technologies.

What, then lies in the other cells? Among other things, economic invest-ments subject to being Ponzi schemes or just long-term failures are found in cell two and depend on moral trust; trust in the strength of bridges etc. are found there too, but depend on material trust.[34] Trust within the institution of science is found in cells 2 and 4. The establishing of new findings and principles through scientific research depends on the kind of trust we value most because it involves willingness to invest work, time, status or material resources in the validity of claims – this is what we will call the pursuit of 'difficult and danger-ous truths'. New findings and principles cannot be demonstrated by replication of experiments because what is at stake is the very principles upon which the experiments depend. The older principles of science are matters of reliance but to establish new ones depends on self-conscious judgements of the integrity of scientists and their work.[35] This will be discussed at length below under the heading of 'Domain discrimination'. An example is the case of 'Checkov'.

That said, the concepts of short-term and long-term are flexible. Even busi-ness transactions where the parties' trustworthiness will become evident in the relatively short-term will be more efficient and run more smoothly if those parties can be relied upon from the outset; this will enable larger transactions to be engaged in with confidence from the outset. We will see when we get to an interview in Chapter 7, with someone engaged in very large-scale business transactions for major international firms, that trust is enormously valued even where any lapse is going to become evident in the fairly short term. Such things are found in the domain of Cell 3.

---

[34] Though, as mentioned in an earlier note, this may depend on moral trust in, say, corrupt societies, where one might have to think about the people who built the bridges.

[35] See Collins (1985/1992).

Examples of trust and reliance can migrate from one cell to another – one kind to another. We have already suggested that routine banking has shifted to the left as more thought has to be put into exactly whether to trust or not. On the other hand, if gamers and hackers keep away, and more and more remote transactions in certain domains are undertaken successfully, perhaps bolstered by the reported experiences of local acquaintances who are already trusted (see Chapter 6, When the remote is really local), calculated trust can turn into something more like reliance and the interactions can migrate to the right. These kinds of shifts are social changes – the nature of society changes when this kind of thing happens – the contents of the 'taken-for-granted' changes. It is becoming clearer nowadays that in so far as at least some members of governments and their advisors were once found in Cell 4, where they could be *relied on* to have the best interests of the whole nation at heart, they are fast migrating to Cell 3, with moral reliance turning into transactional trust aligned with a complex set of interests that are not necessarily those of a considerable proportion of the nation.[36]

Incidentally, in economic terms, shifts from reliance to self-conscious trust are very inefficient because transactions are slowed while trustworthiness is having to be expensively enforced or estimated rather than taken for granted; expensive safeguards are unnecessary where people can simply to rely on each other without thinking. Strangely, much 'Western' mainstream economics, with its stress on financial incentives as the driver for action rather than social cohesion or professional pride, seems to encourage this kind of inefficiency. One consequence for the stress on incentives is the terror of wealthy executives, led to believe that only huge incentives can motivate people to work to the best of their ability, when they enter hospital or the like, where their life and death are in the hands of very low-paid nurses and auxiliary staff and relatively low-paid anaesthetists, doctors, and so on. They must place their very lives in the hands of a very different kind of person to themselves, a person whose salary is not based on the incentive principle! (Or are they terrified? Do they really believe the rationale they present for their own rewards?) Or, consider an event that took place around the time of writing (June 2019): the passion and the huge resources of time, energy and money spent by a million people, among them Collins and his academic son, cheering the Liverpool football team on its victorious return from winning the Champion's League in Madrid in 2019.[37] Professionalism – that idea withering under the attacks of right-wing politicians and their economic outriders – taps into this kind of pride in performance, of

---

[36] Under pluralist democracy, even minority interests are taken into account but that seems less and less to be the case (Collins et al. 2019).

[37] One of the authors (Collins) was in the crowd and noticed that it was made up of complete families, with every member decked out with expensive favours and the kids wearing very expensive team-shirts, and all of them ready to stand in the cold for up to four hours waiting for the team bus to pass for no financial reward at all.

oneself or others, and the motivation costs nothing! Where can we find this in economic theory?[38]

The questions about trust and reliance addressed in much of the rest of the book are about how they are built when technological solutions are not available. The main thesis is, of course, that trust and reliance start with the local and the face to face; even when technological solutions are available, they are grounded in the face to face.

As intimated, where trust and reliance are especially precious, as in the case of building new knowledge in science, where the consequences for failure are severe, we always fall back on the local. Even when we seek a builder or a plumber, we ask for recommendations from our trusted friends and acquaintances.[39] This model of trust and reliance, the remote being parasitical on the local, reflects the deep pattern of communication; there too the remote is also parasitical on the local, because local interaction is what builds languages and societies in the first place. It has to be this way because language depends on tacit understanding and tacit knowledge is transmitted only in face-to-face interaction.

The role of communication and expertise in trust

We must now explain what we mean by communication. The key is a sentiment from the introductory section: anyone who wants to acquire a specialist ability starts by immersing themselves among people who already possess that ability. As intimated, what we mean by communication includes not only the transfer of *information* but the transfer of *implicit understandings*; it is the second that presents all the subtle problems because the first can be handled by books or machines once one has tacitly acquired the means to understand them. Human life starts with socialization, or enculturation, through immersion in the 'bath of language' and continues through further immersions in the worlds, and particularly the languages, of specialist domains. There are two foundational elements to this view, one known as 'studies of expertise and experience', or SEE, and one known as the 'fractal model of society'. A crucial concept that underpins both is 'interactional expertise'.

---

[38] Whilst self-interested utility-maximization is the dominant model of rationality within mainstream economics, alternatives do exist within more heterodox elements of the economics profession. Etzioni's *Moral Dimension*, published in 1988, provides an overview of these views and an argument for an approach to economics that sees economic and other choices as influenced by cultural norms and values and not, therefore, solely determined or explained by individual wants.

[39] Though internet companies desperately look for ways to substitute for these local recommendations with reviews and satisfaction reports.

Studies of expertise and experience (SEE)

The SEE approach, which stands for 'studies of expertise and experience', differs from traditional philosophical or psychological approaches to expertise, and from many sociological approaches to expertise, by defining expertise as membership of an expert social domain. This resolves many of the standard philosophical problems of defining expertise – notably deciding who is an expert when experts disagree or when knowledge changes over time. For SEE, astronomers and astrologers, and chemists and alchemists, are all experts so long as they can, or could, pass as experts among other astronomers, astrologers, chemists or alchemists, respectively. As we will see under the next subheading, ubiquitous abilities, such as speaking a natural language, are also treated as expertises – 'ubiquitous expertises'. 'Passing' as an expert is defined, at least in principle, as passing an Imitation Game; this test of domain-specific conversational competence – a Turing Test where people instead of computers play the role of the intelligent entity – turns on the possession of interactional expertise. Features of expertise understood in this way are: (i) that expertise is something real, not merely attributed by others; (ii) that the reality of expertise is not recognised by its truth or efficacy but the, in principle, testable, fluent membership of a social group; (iii) that the process of becoming an expert and becoming socialized are the same – e.g. a member of a society is an expert in the ways of that society (has acquired what counts as the ubiquitous expertise of that society, however small and esoteric it is) – and becoming such an expert is what enculturation is; and (iv) that expertises may be as ubiquitous as knowing how to live in a national society and speak the native language or as esoteric as, say, gravitational wave physics or even some speciality within gravitational wave physics. Other useful concepts emerge from this approach:

Possession of what we call '*contributory expertise*' – the ability to contribute to a domain as a full-blown expert – is usually acquired through full enculturation into the language and practices of a society or a specialist domain. It is, however, very rare for anyone to be an expert in all the practices of a domain – it might happen in small, technologically undeveloped societies, but most societies and specialized groups operate with a division of labour; this means that even fully socialized individuals have the ability to perform only one or two specialist practices within the collective body of practice of the whole society. That this is the case is hard to notice, and ethnographers sometimes gain kudos by claiming to have been fully engaged in the practice of some sub-group such as boxers. But, to continue with the example, no boxer, professional or ethnographer, has the practical knowledge of boxers in a very different weight-division, or of corner-men, or of boxing managers, or the many other activities associated with boxing.[40]

Interactional expertise involves acquisition of the spoken (and written) language of the domain. This is, in a sense, something less than acquiring practical

---

[40] This argument is developed in Collins and Evans (2016).

abilities though some, primarily, interactional experts, such as technically competent managers, are also contributory experts. All contributory experts will acquire the language of the domain on their way to becoming contributory experts – the language is what directs their attention to what they need to be seeing and doing to acquire a practical expertise. Try to imagine a practical specialist in a domain who knows nothing of that domain except the narrow set of things they are good at practising – one would not consider them an expert in the domain, but an employee doing a paid task. In another sense, then, interactional expertise is something more than contributory expertise; interactional experts may or may not be practitioners within a domain but, whichever, they become fluent in the entire 'practice language' of a domain; if the domain is an esoteric one, they will be specialists when viewed from the perspective of society as a whole. This practice language has to be general enough to enable those who have specialist practical abilities to communicate with each other and make it possible for divided labour to blend into a collective whole. So interactional expertise is more than contributory expertise because it spans an entire domain rather than involving a specialist practice within a narrow subdomain. But, to repeat, interactional expertise is (often) less than full-blown expertise because it does not necessarily involve the acquisition of any of the narrow practical skills of the domain whereas contributory expertise will usually (managers and similar roles aside) involve the mastery of at least one of those skills.

These properties of contributory and interactional expertise are represented in Figure 1.1, which was first drawn to indicate the various practical specialists in the domain of gravitational wave physics. Each specialist's narrow set of practical abilities is represented by his or her individually numbered

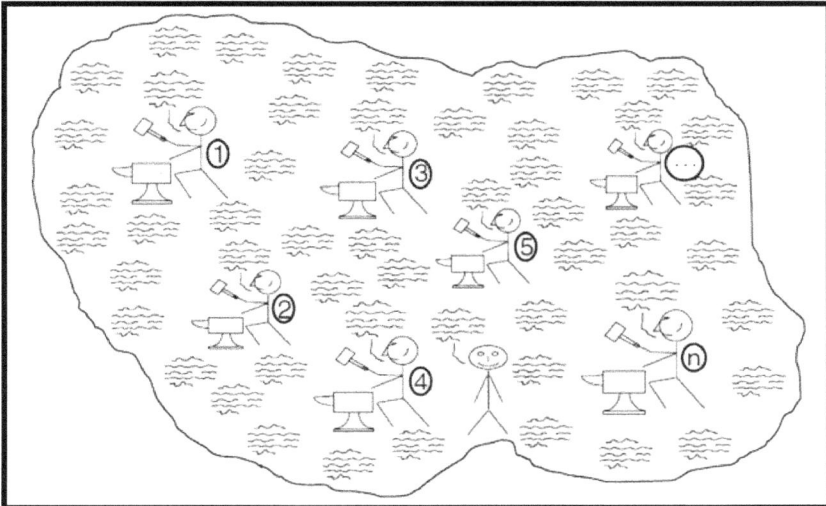

Figure 1.1: A practice domain (originally Collins 2011, Fig. 2).

hammer and anvil, but each is linked through their sharing of the common practice language represented by the identical bundles of waves that flood the whole domain.

There are a couple of additional technical complexities involved in understanding the relationship between interactional and contributory expertise. The first is that, as explained, certain *contributory* roles in a domain may involve only interactional expertise! To repeat, an example is the role of the technical manager of a scientific project who has to grasp the full practical complexities of the whole domain but may not be able actually to practise any of them. A beautiful example is offered by Gary Sanders, the project manager of the 30-metre telescope who is co-author with Collins of a paper on interactional expertise.[41]

> I can sit down in a group of adaptive optics experts who will come to me and say, 'Gary, you're wrong, multi-object adaptive optics will be ready at first light and it will give the following advantages,'—and others will say, 'No, it's multi-conjugative adaptive optics,' and I can give them four reasons why we should go with multi-conjugative adaptive optics based on the kind of science we want to do, the readiness of the technical components, when we need them, and so on. And I will see as I am talking about it that the room is looking back at me and they're saying, 'He does have a line, he's thought it through, and yes.' But if someone said to me, 'OK Sanders, we agree with you, now go and design a multi-conjugative adaptive optics system,' I couldn't do it. I couldn't sit down and write out the equations—But I can draw a diagram of what each part does, where the technological readiness of each one is, what the hard parts are—I know the language. I actually feel qualified to make the decisions. (p. 629)

The second practical point has to do with the subtle interaction between practical and linguistic life and the individual and society. On the one hand, since we came to understand interactional expertise, we know that immersion in the talk of a specialist group alone can convey an understanding of the practices of that group even though the immersed person has not practised. But this applies only to the individual – the practical language of a specialist domain could never have developed in the first place without people practising: without people playing tennis there would be no tennis language so there would be no interactional experts on tennis. For philosophers, this is how we square Wittgenstein's idea that concepts and practices are but two sides of the same coin with the idea of interactional expertise. It is also why we will be able to claim when we get around to discussing, for example, 'Winograd schemas' (see Chapter 2, Second language learning), that a proper understanding of language involves

---

[41] Collins and Sanders (2007).

an understanding of practical life: true fluency in language acquisition, such as by deep learning techniques, is going to have to involve the kind of immersion in everyday language which only those with a practical understanding of life engage in, and that level of immersion is still a long way off for machines and computers. Furthermore, it seems that, at the moment, only humans are capable of the creative rule-breaking that is characteristic of fluent language use.[42]

Interactional expertise is often mistaken for something far less than it is – sometimes those who misunderstand the idea or want to criticize it say it is mere 'talking the talk' rather than 'walking the walk'. But this is a serious misunderstanding. Interactional expertise is the facilitator of contributory expertise, enables the division of labour and therefore makes it possible for practical experts to specialize, and, as has been intimated, it has a broader reach. The only thing missing from interactional expertise is the ability to practise in one (or two) narrow specialisms within the specialist domain, but one can still make the technical judgements that involve a practical understanding of the domain.

The Imitation Game can be used to demonstrate what is meant by being able to make technical judgements in a specialist domain and to define what is meant by a specialist domain. The Imitation Game is the forerunner of the Turing Test, in which a judge interrogates a hidden computer and a hidden human. Turing suggested in his famous 1950 paper that if the judge could not tell which was which, the computer would have to count as intelligent. He took the idea from a game in which men tried to pass as women or vice versa. This has subsequently been adapted into a method for sociological research. To indicate the carefully worked out method we capitalize, thus: 'Imitation Game'. In the Imitation Game, we have non-experts in a specialist domain trying to pass as experts when the Pretender and the genuine expert (Non-Pretender), are interrogated by an expert Judge. A specialist domain is one where it is possible for the Pretender to fail to pass because to pass they would have to acquire some specialist understanding – that is, in a specialist domain there has to be a body of understanding that one can fail to acquire. The Imitation Game has been deployed to show that in such domains, interactional experts have a level of technical understanding that is indistinguishable from that of contributory experts in written tests. Interactional expertise is not merely 'talking the talk'; it is better described as 'walking the talk'.[43]

---

[42] See, e.g., Collins (2018).

[43] There is now a large literature on interactional expertise. For an early explanation of the idea see Collins (2004a); Collins and Sanders (2007); with Collins and Evans (2015) being a later history and overview. For Imitation Game research see Giles (2006); Collins et al. (2006); Evans and Crocker (2013); Collins and Evans (2014); Collins (2016 or 2017); Collins et al. (2017); Wehrens and Walters (2017); Arminen et al. (2018); Evans et al. (2019).

The fractal model of society

Young children, as they are brought up, undergo socialization. Through sociali-zation, children acquire fluency in their natural language. They also learn what counts as true in their societies, and the right way to behave – what counts as clean and dirty, good manners and bad, how to physically interact with others and so on – they learn to *rely* on some things and not rely on others. We will call this 'primary socialization'. 'Secondary socialization' is what is involved when humans start to acquire the particular skills involved in specialist occupations, second languages, and so on. SEE, as explained, treats all kinds of socialization, whether into a specialist skill, or a language or the way-of-being-in-the-world that characterizes a society, as essentially the same. This resolves the problem of what counts as a society or counts as a group or specialist domain: it is anything into which you can be socialized or fail to be socialized, and is not anything into which you cannot become socialized or cannot fail to become socialized (such as wearing laces in your shoes or being brown haired). Things into which you can become socialized or fail to become socialized are found at every scale from whole societies to small groups such as cricketers, Christians, gravitational wave physicists, gravitational wave-form calculators and so on: these smaller groups are not normally thought of as societies but SEE treats them as mini-societies. The point is that all these groups, or whatever you want to call them, are similar in that you become members of them – that is people who can pass as members of them when interrogated by other members of them – through socialization. And this means that you understand their respective practice languages flu-ently enough to know when a rule of the language is being followed, when it is being broken and, indicating true fluency, when breaking a rule indicates that a mistake is being made and when breaking a rule is a creative act. 'Disguster-ous', and 'snozzcumber', even though my computer has flagged them as spell-ing errors, are examples of Roald Dahl being acceptably creative; 'strorberry', and 'poodel' are rightly-flagged as mistakes.[44] As far as the project of this book is concerned, however, there is a difference between primary and secondary socialization. Babies and young children have fewer communicative resources to draw on than adults; babies and young children cannot use the telephone or email or use social media or experience virtual reality. Therefore, showing that face-to-face interaction is irreplaceable as part of the process of socializa-tion, even as remote communication's bandwidth increases, is harder in the case of secondary socialization than primary. Our main topic is the harder one – secondary socialization.

Societies, groups, or domains into which it is possible to be socialized can, then, be not only big or small, but also esoteric or everyday. Furthermore, they are embedded within one another, and they overlap in multidimensional ways.

---

[44] See Collins (2018) for this difference and its significance being worked out in greater detail.

It is their similarity in respect of the crucial features of socialization and fluency that leads to the fractal model of society – a fractal being a structure which retains the same form at every scale. Examples are a cauliflower, which has florets within florets within florets all the way down until we reach individual cells, or the coast of Norway which has fiords with mini-inlets of the same form within the fiords, and mini-inlets within them, and so on all the way down until we reach grains of sand.

The fractal idea is the best way to understand the way that specialist domains cascade. Thus, there are specialist practical domains within, say, gravitational wave physics – represented by the hammers and anvils of Figure 1.1 all linked by a common language, but a language which is itself special to gravitational wave physics. But each of those specialist practical domains can be seen as a self-contained group and it will have a mini-language of its own, suitable for talking of, say, how to glue mirror-supports to mirrors, or how to polish mirror surfaces, or how to calculate the waveforms of inspiralling black holes. These mini-languages will not form part of every gravitational wave physicist's fluency. The fractal idea enables one to switch between levels of specialization without confusion.

Figure 1.2 is a simplified diagram of UK society intended to illustrate the fractal model. It is simplified: a true society would have to be represented in many dimensions because of the many ways it can be sub-divided into different groups.[45] Primary socialization is responsible for much of the upper oval – notably natural language acquisition; this is the domain of socialization into one's native society. Some other aspects of this general socialization are, however, matters of secondary socialization. Most of the rest of the domains are matters of secondary socialization alone, though some aspects of parenthood are learned in the family home.

The bottommost oval – 'dwile flonkers' – is there to show where the fractal ends in the same way as the cauliflower and the Norway coast fractals end in a few cells and a few grains of sand, respectively. Dwile flonkers are collections of humans who come together to play a jokey game. They are probably *not* a group in that reading the Wikipedia entry on dwile flonking is almost certainly enough to be able to pass as a dwile flonker in an Imitation Game – there is no real understanding or socialization involved. If this is true, Dwile flonkers are, in respect of the fact that you cannot succeed or fail in becoming a dwile flonker, like the already-mentioned people who wear laces in their shoes or

---

[45] Abbott (2001) offers a similar argument about the fractal nature of specialist domains in his book *Chaos of Disciplines*. The argument we are presenting here is compatible with Abbott's, but is focused on culture and knowledge more generally, not only within academic disciplines. This more general work builds on ideas introduced in previous work, such as Collins and Evans (2017b).

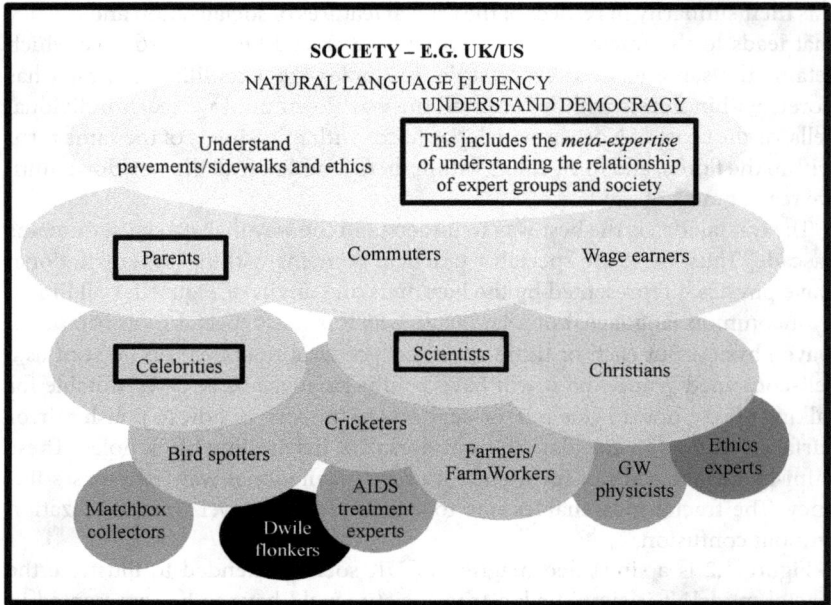

Figure 1.2: The fractal model of society.

have brown hair. To repeat, *groups*, proper, involve socialization, understanding, shared fluency in language and a potential test which could be failed.

Because all kinds of socialization are the same at one level, we can get *some* indication of what is needed for secondary socialization to succeed by looking at the experience of ethnographers, anthropologists and sociologists as they acquire interactional expertise.[46] Secondary socialization certainly seems easier than primary socialization. Primary socialization involves contact between parent and child for most of the time that the child is awake. But secondary socialization can be accomplished by meeting an initially unfamiliar group's members for a few days every now and again, and this is a very small proportion of daily life and almost certainly a very small proportion of the time over which this socialization takes place, even allowing for any follow-up conversations and emails that might occur. The much larger amount of contact that it is mistakenly thought to be required for thorough socialization in sub-groups probably goes back to the ideals of the early anthropologists who immersed themselves in the lives of distant peoples for years (two years among the Trobriand islanders for Malinowski). But modern ethnographers also set out

---

[46] One or two try to gain practical skills too, but we are not concerned with practical virtuosity or the development of practice-based cultures but the understanding of a culture, which comes from acquisition of fluency in the domain language.

to understand the people they investigate, the standard, according to accomplished ethnographer, Gary Alan Fine, being 'saturation' – the point where nothing new is being learned from extended contact. Fine, who has completed about a dozen ethnographies, writes:

> For my Little League baseball research, it was three years (summers: about two months each), being with the boys for 3 afternoons about 4 hours each ... The Dungeons and Dragons research was 18 months, two nights/week (4–8 hours each). Restaurants; 4 restaurants, 5 days/week – 4–8 hours; Meteorology 18 months – 3–4 days/week (2–10 hours). ... (personal communication with Collins, 26 November 2018)

Given Fine's prominence in the field, we can take this as reasonable for an ethnographer, but we should note that it is far from the expectations of the early anthropologists. It is, however, probably pretty similar to the proportion of time spent in the company of their respective groups by the participants themselves – the baseball players, gamers, restauranteurs and meteorologists. So, this provides a suggestive requirement both for ethnography and secondary socialization in general, and it is far less than the contact time of parents and young children. Of course, to reach the higher level of practical accomplishment which might lead to fame and fortune, or even a secure professional salary, is another matter precisely because such persons are aiming to stand out from other members of the field, not acquire its culture.

A relatively young anthropologist answered the same kind of question. Olaf Zenker said:

> research on Irish identity and language revivalism in Catholic West Belfast lasted 15 months, mostly in West Belfast, but there were also several weeklong Irish language classes in Gaelic colleges in Gaeltacht areas. I conducted participant observation for about 3–4 hours per day in various Irish language contexts such as language classes, Irish-language bars, cafes and restaurants, in the Irish language cultural centre in West Belfast, Irish-medium schools, within informal settings of Irish-speakers in their private places etc., in addition to conducting 145 semi-structured formal interviews. There were some instances of daylong fieldwork, but these were combined with more limited days and some void days giving the average of about 3–4 hours of ethnographic fieldwork in a day (personal communication with Collins, 5 Dec 2018).

Kennedy reports that his work with wildfire managers started with a shorter but more intensive period of immersion. The primary immersion was concentrated into a period of eight months, with 160 days of contact ranging from 3 to 15 hours per day. Some of this was done via some 200 interviews, while the majority involved fly-on-the-wall and participatory observation in headquarters and in remote sites. The biggest challenge was the fractal model: clear cultural distinctions emerged between wildfire managers in different provinces, allowing for comparative reflection during the process, but also resulting in subtly different experiences of socialization from case to case. This perhaps explains his relatively long period of immersion since he was really studying a number of slightly but significantly different communities.

Collins's engagement in gravitational wave physics was less intense than that of these others, but its duration was 45 years with a central ten years that was the most important in terms of becoming socialized. Collins began with a series of interviews in 1972 and would now claim that, more important than the number of hours of contact, is to know what you are trying to accomplish; one can certainly point to instances of intensive contact with a group where nothing much in the way of understanding was accomplished because this was not the aim, whereas one can begin to understand the mini-society from even a few interviews if this is their self-conscious purpose.[47] The longest time Collins spent with respondents in any one place was less than a month, but the period he counts as his most intensive, which ran from the mid-1990s to the mid-2000s, involved visits to conferences and instrument-sites all over the world but of no more than four or five full days and evenings at a time; there were, perhaps, half-a-dozen of these a year. This amounted to no more than a month or two a year in total but stretched over the better part of ten years. Engaging in talk in formal and informal settings in these places over this period was enough for him to come to understand the field well enough to pass as a gravitational wave physicist in two Imitation Games ten years apart, where gravitational wave physicists, judging in Turing Test-like circumstance, failed to distinguish between him and one or more other professional gravitational wave physicists – a very high level of understanding.[48] The first of these tests was passed after about five years of this kind of immersion in the field.

Language is central to understanding even practical domains and language is central to personhood; these accounts of how ethnographers and sociologists gain interactional expertise tell us how the foundation of socialization is acquired. We will come back to the importance of language again and again, all the way from the difference between people and cats through to the importance of scientific conferences.

Transactions, socialization and trust

In Table 1.2: Trust and reliance in short-term and long-term, we set out some kinds of transaction, noting that we were more concerned with cells 2, 3 and

---

[47] In Collins (1984 and 2019), the case of Festinger et al. (1956) is cited as such a case while the claim that knowing what you are trying to do is the most important thing is developed.

[48] Giles (2006); Collins (2017, Ch. 14).

4 than cell 1, where transactional trustworthiness can be assessed in the short term, perhaps with the aid of technological means. Returning to the theme, we are also mainly concerned with transactions where the parties are deeply concerned with the integrity of the outcome. There are kinds of financial transaction where the outcome is not clear until we get to the long term. Ponzi schemes are one example; confidence tricks are another example – in both cases a façade of short-term trustworthiness is put in place but it is a sham. The hugely financially successful sales of sub-prime mortgages bundled into derivatives is another example – the banks acting, unnoticed by most, in totally untrustworthy ways until the 2008 crash. Another massively financially successful scheme that depends in no way on trust was, or is, high frequency stock-trading – which is a legal technological wheeze, made possible by remote communication, to make lots of money at the expense of the mass of ordinary investors.[49] In our argument about the irreplaceability of the face to face, what we are concerned with here is not financial success but the kind of transaction where intrinsic rewards – the reward of doing a good job – is the principle motivator, because, in such cases, integrity is built in: other kinds of rewards, should they accrue, are secondary. We have already mentioned the professions as examples of groups primarily motivated by the successful achievement of a goal – intrinsic rewards – rather than an extrinsic return such as money. The search for scientific truth is one example and the search for gravitational waves is a case we know well.[50]

It was 100 years from Einstein's theoretical derivation of the existence of the very weak gravitational radiation to the detection in 2015 of such a wave, with experimental attempts to detect the waves beginning in the 1960s, and eventually costing about a billion dollars. During this 50-year period, half-a-dozen claims to have seen 'stellar origin' waves were made but all were refuted *by the scientific community itself*, driven by the extremely improbable energy requirements implied. For example, one claimed sighting, related to a supernova – an exploding star – seemed to require the, impossible, conversion of matter with a mass of 2,400 of our Suns into energy, and this, literally, *incredible* requirement was pointed out in the paper itself. The 2015 discovery, which depended on instruments so sensitive that the energy requirements were not unreasonable (the conversion of three solar masses into energy consequent on the merger of two black holes), was kept secret for five months as the more than 1,000-strong community investigated every possible mistake that could have been made

---

[49] Lewis (2010 and 2014) for further description of the two cases. For high-frequency trading, https://www.lrb.co.uk/v41/n05/donald-mackenzie/just-how-fast is informative.

[50] Of course, when other values, such as passion for the Nobel Prize, take over, things can go wrong. Such seems to have been the case in the claim that primordial gravitational waves had been detected by a telescope called BICEP2 (Keating 2018).

(and the final vestiges of nervousness were not dissipated until a second, unannounced, sighting was made). During this period the scientists' constant companions were thoughts about previous examples of claims that had not been supported, both of stellar-origin gravitational waves, and other phenomena such as gravitational waves of primordial origin and the well-known 'discovery' of a magnetic monopole of which a second example has never been seen. The financial rewards to the discoverers and non-discoverers are roughly similar – continued posts in universities for most, with prizes for a very few of the 1,000+ – but the true reward was to have lived through and been part of the discovery even though, for almost everyone, it left the pay-packet pretty well unchanged. How else could 50 years of experimental effort, mostly fruitless, and often led by pioneers who expected to be dead before there was any success, be maintained? Indicative is the reaction of Collins himself, who wrote sociologically driven accounts of the 50 years of experimental work and of the discovery itself in which he admitted that, from a sociologist's point of view, it would have been better in some ways if the discovery had been more disputable and disputed, but that being associated with the discovery as it was, was one of the greatest and most worthwhile experiences of his life, even though he gained nothing from it except intrinsic rewards.[51]

To illustrate what we are getting at in this section we would need to add a third dimension to Table 1.2 which would represent the importance of the integrity of the transaction to the actors. By changing to diagrammatic convention, we can illustrate three dimensions – this is shown in Figure 1.3.

In the case of the detection of gravitational waves, the criteria of success were indefinitely long term since no-one could be sure the waves would ever be detected, or even if there were any waves to be detected (even Einstein later claimed he was wrong about them), while, to the actors, the importance of integrity was maximal. But the trust relations included both explicit trust – for instance on whether someone's report of a result could be trusted – and implicit reliance – for instance on the wider understanding of the physical world within which this particular project was set. Therefore, the position of the actors in this case is roughly represented by the grey rectangle along the back upper edge of the three-dimensional space.

The detection of gravitational waves case is an ideal, but other transactions may have some of the same qualities. We have already remarked that many transactions with relatively well-defined and fairly short-term criteria of success (bottom of the Y-axis) are still carried out more efficiently if there can be strong trust between the parties (origin of the X-axis). Such a position is

---

[51] The discovery has been disputed (e.g. see *New Scientist*, 3 November 2018, front cover and article by Michael Brooks entitled 'Wave Goodbye', pp. 28–32) but not in a way that is credible enough to be sociologically interesting. Collins's account of the discovery, and some of the doubts, can be found in his 2017 *Gravity's Kiss*; royalties on such books are negligible.

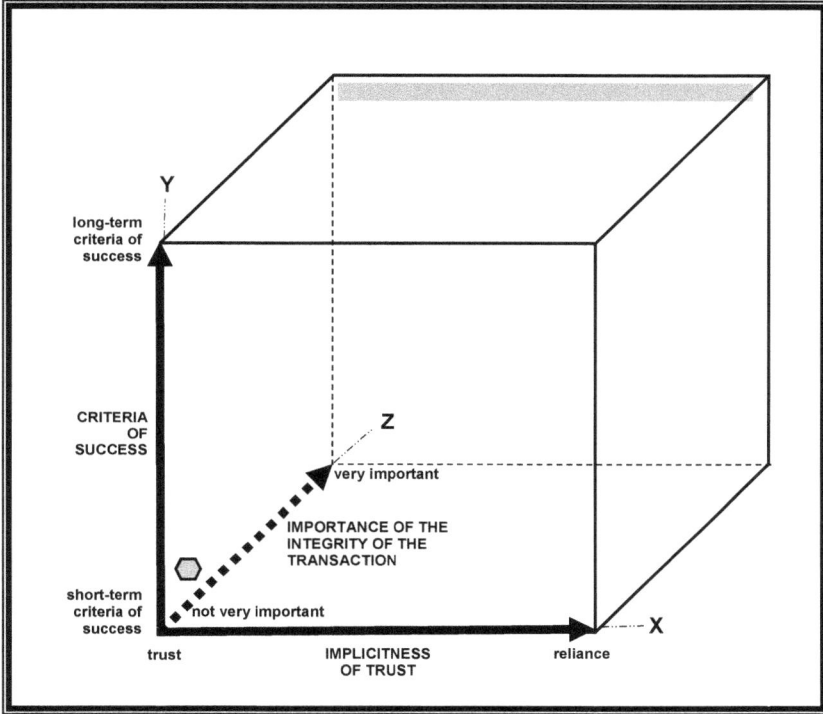

Figure 1.3: Communication space.

roughly represented by the position of the grey hexagon: it is on the left-hand wall because it is trust that is at stake but it is nearer the origin than the end of the Z-axis because nothing of world-changing importance is going on – nothing that will make anyone feel fulfilled as they reflect on their life while lying on their death-bed. The location of blockchain technology (see Chapter 4: Remote Technology and Trust) seems to be the same – if the circumstances are right it can replace efficient, interpersonal trust in the case of those kinds of transaction – which is what makes it a striking technological innovation. The other technological developments, such as reviews on the internet by multiple consumers, are a less reliable version of the same kind of idea.

In what follows, we dive more deeply into these questions of socialization and trust. We explore primary and secondary socialization, the bath of words and the role of tacit knowledge. This, in turn, sets us up for Chapter 3 on face-to-face communication, where the implications of tacit knowledge and socialization will become clear.

CHAPTER 2

Forming Societies and Learning to Trust and to Rely

We now start on a more systematic exposition of Table 1.1. We start with the first line. This refers to the commitment and energy that is generated by people being willing to expend their time and resources in travelling to meetings, demonstrating that the joint goal in prospect is being taken seriously.

We will deal with this again when we get to a 'reflexive look' at our own project in the next chapter, but, for now, it is worth pointing out that this very project is an example of the energy that co-presence can provide. After the grant application for the initial version of this project was rejected, vital energy and affirmation was supplied by the three or four participants who responded immediately and positively to the question of whether we still wanted to carry on, especially those from abroad who agreed that they were ready to find the resources to come to an early face-to-face meeting in the UK. That the meeting did happen signified commitment by most of the potential co-researchers: the more people are willing to put in resources, the more they affirm the value of what is going on. This is one of the ways new thinking turns into new facts – it is the beginning of the way things are 'socially constructed' – in this case almost literally.[52] This is part of the social construction of knowledge – it is the beginning of the social construction of a new feature of the fractal model of one's society. Barring wrecking disagreements, face-to-face communication avoids the negative inputs that correspond to the energy-sapping delays that various forms of remote communication encourage.

---

[52] Collins himself notes the trust that he was able to generate among gravitational wave scientists by showing his commitment through attending every meeting over a period of years (see below).

How to cite this book chapter:
Collins, H., et al. 2022. *The Face-to-Face Principle: Science, Trust, Democracy and the Internet*. Pp. 45–66. Cardiff: Cardiff University Press. DOI: https://doi.org/10.18573/book7.c. Licence: CC-BY-NC-ND 4.0

Primary socialization

Now we move on to more general aspects of socialization. How do babies learn their language and way of life from the societies in which they are embedded? We know it happens through socialization – immersion in the language and social life of the group whose way of being in the world and way of speaking is to be acquired. But we do not know the mechanism of socialization in any detail, though we know how to manage it both in terms of babies learning to live in their native society, including acquiring fluency in their native language, and in terms of sociologists and anthropologists acquiring an understanding of a new society; in both cases the process is immersion in the society or group.[53] When we say that we do not know the mechanism in any detail, we mean that it is, or certainly was, possible for social psychologists and linguists to argue about whether it was so hard for a child learn a native language that there must be some kind of innate 'generative grammar' in the brain that formed a framework to which any particular language need only be attached – a far simpler task than learning everything from scratch. This is the idea introduced by Chomsky, but it is coming more and more under attack. The arguments in the academic community about how language is first acquired are not over, however, and this shows how little we are certain about.[54] We are going to have to speculate and use our imaginations.

It seems impossible to bring up a child by remote communication without physical contact with parents and other humans. If John Bowlby's *Child Care and the Growth of Love* thesis is correct, a child growing up feeling insufficiently loved would not develop normal affective relations with other humans. At first sight it certainly seems that remote care without human contact would not be enough to convey that love and that there are psychological and emotional necessities that come only with human contact.[55] It also seems almost

---

[53] Collins (2019).

[54] For a forthright attack on the Chomsky view see Reber (2011). Reber's view is that his mechanism of 'implicit learning' accounts for language learning. Reber, in experiments in which he exposed human respondents to meaningless sentences deploying an artificial grammar that he had invented, showed that humans can grasp novel grammars 'implicitly' without even knowing that they were learning such a thing. This is why Reber's work, even though it uses laboratory experiments in artificial situations, relates to Collins's on tacit knowledge, which deploys natural situations – namely the transfer of knowledge between scientists. But even Reber does not solve the problem of how these patterns are acquired except in terms of some generalised neural-net model of learning.

[55] Whether robots could replace human contact is not our topic: our topic is the remote. But our discussion of language, a little later in this paragraph, does intimate that, for robots to be successful in teaching human-style

certain that fluent native language acquisition, which is effectively culture acquisition, depends on face-to-face immersion in the bath of language with its unexplicated but subtly informative distribution of words, silences and body language. Could a child grow up to be a fully fluent language speaker if the bath of language in which it was immersed as it developed was not accompanied by the real touches and other subtly reinforcing physical sensations that accompany a human upbringing? Suppose the child experienced simulations of those sensations – would that work? Or is it that families work as effective language teachers because there are haphazard encounters between their members driven partly by the infant instinctively demanding attention and interaction in order to learn? That kind of thing seems hard to reproduce outside of a small group of humans.

There is a literature which supports the view taken here. It seems that real fluency – especially native accent – can be achieved only if 'immersion learning' – that is learning among native speakers – is begun before a certain age. For instance, a study that shows that infants exposed to Mandarin as a new language picked up the sounds rapidly if taught in a socially interactive context with eye-contact and touching, but learned nothing if exposed to video film of the same stimuli.

> Social influences are important in speech learning. Infants learn more easily from interactions with human beings speaking another language than they do from audiovisual exposure to the same language material, and their speech is strongly influenced by the response of others around them, such as their mothers. The importance of social input in language learning has some similarities to social influences on song learning in birds. (Kuhl 2004, p. 831)

> Social interaction creates a vastly different learning situation, one in which additional factors introduced by a social context influence learning. [This] could operate by increasing: (1) attention and/ or arousal, (2) information, (3) a sense of relationship, and/or (4) activation of brain mechanisms linking perception and action … the live learning situation allowed the infants and tutors to interact, and this added contingent and reciprocal social behaviors that increased information that could foster learning. During live exposure, tutors focused their visual gaze on pictures in the books or on the toys as they spoke, and the infants' gaze tended to follow the speaker's gaze, as previously observed in social learning studies … Referential information is present in both the live and televised conditions, but it is more difficult to pick up via television,

---

fluency in language as well as emotional security, they would have to be the facsimiles of humans that are regularly found in science fiction but not in prospect in the actual world of AI (see also Collins 2018).

and is totally absent during audio-only presentations. Gaze following is a significant predictor of receptive vocabulary. (Kuhl 2010, p. 720)

Video learning, we can surmise, not only removes all the guiding touches and encouraging hugs but also, inevitably, strips off the richness of body language and gaze.[56]

Second language learning

One might have hoped that the whole problem of the importance of face-to-face interaction could be resolved 'at a stroke' by studies of *second* language learning, corresponding to our secondary socialization, since the conventional wisdom is that immersion among native speakers is vital for fluency. How does efficiency of second language learning compare with first language learning and how does learning a second language with immersion compare with learning without immersion? There is indeed an interesting literature on these matters but it turns out not to answer our kind of question because its criterion for success in language acquisition is grammatical accuracy rather than fluency (see Appendix 5), but grammatical accuracy is only a contributor to fluency and one that is not too difficult to reproduce mechanically as grammar checking functions in word processing software reveals.

Perhaps it is because of the influence of Chomsky that language learning has been thought of as a grammatical puzzle rather that the acquisition of a skill or expertise. For our purposes we have to look elsewhere to see why immersion might be better than intense distance learning for language fluency and use a different kind of criterion for the acquisition of fluency.[57] Really demanding fluency depends on understanding the way a society works, not just understanding grammatical rules – or, more exactly, that understanding the grammatical rules of a native language is interwoven with understanding the society in which that native language is spoken, at least if the aspiration is to achieve the level of interactional expertise. A neat illustration of this can be found in the debate about artificial intelligence, though it is surprisingly little known even in that narrow domain. Terry Winograd, the AI pioneer, invented, or discovered what have become known as 'Winograd schemas' and explained them in his 1975 PhD thesis.[58] A Winograd schema is a sentence like the following:

'The trophy would not fit in the suitcase because it was too big/small'.

If translated into a language with gendered nouns, such as French, and if the final word is 'big', the 'it' should be translated as, 'il', since it refers to 'le

---

[56] Eckmann (2009) shows how difficult it is likely to be to capture almost subliminal 'tells' via video link. See also Manstead, Lea and Goh 2011.

[57] For a paper which draws the contrast and proposes a change see Chater and Christiansen (2018).

[58] They are discussed at greater length than here in Collins (2018, Ch. 10).

trophée'. But if the final word is 'small', the 'it' should be feminine – 'elle' – since it then refers to the suitcase – 'la valise'. Thus, to get the grammar right in this translation one needs to understand the world to which it refers – in this case the world of trophies and suitcases – and understanding the world is best done by living in it since one never knows what one will have to understand next.[59]

Another example of the need to understand the world, invented by Collins, is this simple sentence: 'I am going to misspell wierd so as to make the problem of fluency obvious'; spell checkers always flag 'weird' whereas a skilled copyeditor will leave it uncorrected.[60] A second example, invented by Collins, is the following English sentence in need of correction:[61]

> Gofers sumtimes where +4s. One might think they would ware +fores. Could it be that half a pair of +4s is a pair of +toos and that two pears of +4s are +8s. If that's right I'll eat my pear. What's odd about it is that a plus four plus a plus four is usually ate.

In this instance, rectifying the English requires knowledge of golfers and their garments as well as arithmetic, spelling, fruit, diet, homophony and the phrase, 'I'll eat my hat'. One can invent any number of such examples in an open-ended way.

This combination of grammar and cultural understanding is better tested with a demanding and inventive Turing-type test or 'imitation game' than with pre-set grammatical exercises. In practical terms, the Imitation Game (capitalised), can be used since it is a Turing Test but with human participants. This, rather than grammatical accuracy, is the definition of fluency that underpins the arguments in this book.

Going back to primary socialization, even in science fiction some, at least, find it hard to imagine it being accomplished remotely. In the 1950s, Isaac Asimov published a story, *The Naked Sun*, about a planet where everyone communicates via 'trimensional' viewing – extremely high-fidelity video. The location of Asimov's story is 'Solaria', an Earth-sized planet with a population limited to 20,000, separated individuals living on vast estates evenly distributed across the globe. Each estate is served by 10,000 robots with 'positronic brains', keeping the lone person fed and productive. The individuals have become used to their 'isolation' and find it normal. They are separated by huge distances nearly all the time, communicating 'trimensionally'. The bandwidth is so high that 'Baley', a detective visiting from Earth to investigate a murder, does not at first realize he is not conversing face to face with a Solarian. Only when his conversational

---

[59] Though the test is not quite so neat, Gricean 'implicatur' requires an understanding of the world in the same way (see Davis 2019).

[60] This example taken from Collins (2018) wherein there is extensive discussion of this kind of demanding Turing Test problem.

[61] Collins (1992, p. 737).

partner breaks the connection and instantly disappears, along with his chair and the walls behind him and floor below him, does Bayley realize that the other person was never really there. Only after some practice does Baley learn to recognise remote trimensionality by looking out for joins between one pattern of walls and flooring and another.

Asimov invents an entire culture around trimensional 'viewing'; Solarians have a horror of localised interaction, which they refer to as 'seeing'. They consider *viewing* to be natural and wholesome while *seeing* is rude and dirty: 'Imagine – I might be able to smell you, or breathe in the same air you have just breathed out!' Since procreation does still involve physical contact, 'children' is a dirty word, as Baley discovers when he asks a shocked and embarrassed Solarian if she has offspring. Procreation is strictly limited so as to maintain the population at 20,000 and the duty of the necessary physical contact assigned, and then only rarely, to certain couples after genetic matching. But even Asimov could not imagine human life in the nursery being founded on trimensional viewing, extraordinarily advanced as it is.[62] Crucially, in Asimov's story, in spite of the horror of direct contact, babies are taken early from parents and brought up together in farms where they play vigorously together, with the viewing culture not displacing this physical interaction until the early teenage years. Asimov, like us, cannot imagine, even in a society like Solaria, with a dislike of physical contact and with no shortage of intelligent robots, that early socialization will not need physical interaction.[63]

Asimov's story warns us, *inter alia*, that technological change brings organizational and cultural and change. The concern in this book could be said to be that we are gradually shifting away from a 'seeing' culture toward a 'viewing' culture while assuming our culture and society will remain the same in all other respects.

---

[62] Asimov's book *The Naked Sun* is listed in the bibliography as '1991' but it was first published in the mid-1950s.

[63] It may seem odd to cite a science fiction novel at the outset of our analysis. We are not the slightest bit interested in 'the plot' of *The Naked Sun*, which has Baley solve the murder in part through exploiting the tension between viewing and seeing, but in developing the context for the plot, Asimov is really being an (unwitting?) anthropologist of our own society – exploring how *our* communications with each other work. It has been argued that Wittgenstein's 'philosophical investigations' are similarly, unwitting sociology, being rooted in the way we, in our society, think and act – even to the point of little stories about imaginary sellers of wood and so forth (e.g. Bloor 1983). We might think of these little stories, along with Asimov's imagined society, as sociological 'thought experiments'. Then again, think of actual societies, such as the Azande with their extraordinary poison oracle, about whom philosophers and sociologists debate endlessly (e.g. Wilson 1970). Given what most of us actually know about the Azande, a thought experiment would serve just as well!

We are not experts on primary socialization, but, fortunately, our main topic is secondary socialization, about which we know more.

Secondary socialization

As explained, secondary socialization is what happens lower down the fractal which represents society, and it happens as children begin to turn into adults and encounter more domains which provide them with more specialized and local sets of skills and understandings. In primary socialization, language and practice are harder to pull apart. In the worlds of babies and young children there is an indistinguishable mixture of learning to do things and say things, and every young person is acquiring roughly the same knowledge as every other in the same general locality or, in many cases, the entire society; it is *ubiquitous expertises* that are being acquired. In the same way, in the world of babies and the world of young children, there are no specialists in spoken discourse who do not practise, such as the equivalent of technical managers or sociologists and anthropologists; these language specialists, physically challenged persons apart, start to emerge only later in adult life when specialities emerge with division of labour.

When we get to secondary socialization, new kinds of question arise. How can one learn to understand practical things merely by talking about them without doing them? It would not be that strange if knowledge was a collection of facts, but we know that facts rest on a body of tacit understanding that cannot be conveyed explicitly, and some of this understanding is of a highly practical nature. Some of it certainly is conveyed by conversation, as we can recognise with Imitation Game tests for interactional expertise. How, then, is knowledge conveyed in words when it cannot be explained? Experiments on implicit learning (see note 54) show that humans can unknowingly extract grammatical patterns from strings of 'words', but much more than this has to be extracted if interactional expertise is to be explained. Interactional expertise captures everything that can be said about the practices it refers to but, obviously, without needing everything to have been said – which would be impossible; the interactional expert has to be able to say things they have never heard and understand things which have never been explained. Let us begin with some more imagination.

Duck-rabbits and the bath of words

Let us invent a 'just so' story to try to understand how the bath of words can shape perception. Suppose you have two specialized domains in each of which there are many representations of what we, who are familiar with both ducks and rabbits, would call 'duck-rabbits' – see Figure 2.1. These are the well-known 'gestalt-switch' images that can be seen as either a duck or a rabbit but not both

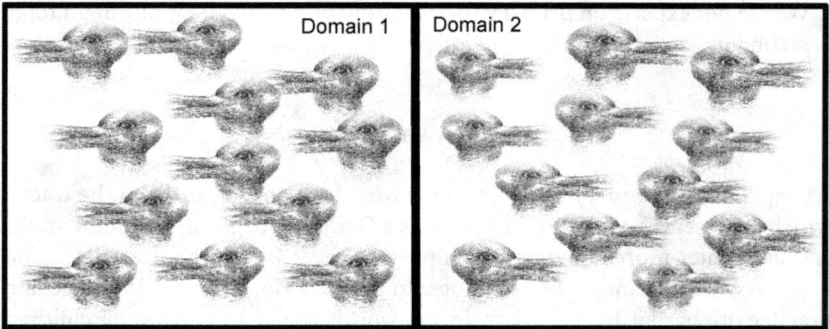

Figure 2.1: Two domains with, as we see it, lots of duck-rabbits in each.

at once. The difference between the domains exhibits itself in the languages, practices and the corresponding perceptions.

In Domain 1, people are always talking about ducks – they love feathered creatures that swim on the water and quack. Children's stories are full of such creatures and popular speech is loaded with references to ducks, both real and metaphorical. There are also duck farms and ducks' eggs and roast ducks and duck-feather quilts. Furred herbivores that live in burrows are unknown, however; there is rabbit-silence. In Domain 2, the opposite is the case – feathered quackers are unknown and unspoken while furred herbivores that live in burrows are an everyday feature of talk, love and diet from baby-hood. What do people from Domain 1 see when they look at the left-hand panel? They see, not duck-rabbits, but a dozen ducks looking left; there are no rabbits and, by extension, no duck-rabbits in Domain 1. What do people from Domain 2 see when they look at the right-hand panel? They see a dozen rabbits looking left; once more there are no duck-rabbits in Domain 2, only rabbits. And this outcome could be predicted from analysing the distribution of words in the two domains because in Domain 1 there will be lots of occurrences of 'duck' and silence when it comes to rabbits; in Domain 2, there will be lots of occurrences of 'rabbit' and silence when it comes to ducks.

Now, within the domains it never occurs to anyone to measure the relative frequency of use of 'duck' and 'rabbit': it is us who are doing this from our external analytic viewpoint. It would never occur to anyone in Domain 1 to note the absence of the term 'rabbit' from their discourse nor rabbits from their lives since they do not know either the word or the creature in the first place – they cannot be inquisitive about their absence; it would be like asking why the term 'quoggle' is absent from our discourse.[64] The same holds for Domain 2 but

---

[64] Note how naturally, you dear reader, recognised, 'quoggle', as a piece of crea-tive rule-breaking whereas a spell-checker flags it, annoyingly, as a mistake. Our copyeditor will understand immediately.

it is the absence of the term 'duck' from the discourse that would never occur to anyone to inquire about.[65] So it is silences as well as sounds in the language from which the inhabitants of these domains are acquiring the immensely practical knowledge of how to see the world; from our god-like viewpoint we see identical representations in both domains and we can say that they are seen quite differently in them and this difference is tied to the local languages. This must be part of the way interactional expertise works and part of the way tacit knowledge is transferred: learn the language of domain 1 and live there and you will see ducks not duck-rabbits; learn the language of domain 2 and you will see rabbits not duck-rabbits. This helps us see how it is that language plays such a crucial role in our way of being in the world.

Humans, sociologists, psychologists and what it is to be a cat

The just-so story of the duck-rabbits has its counterparts in the real world even if they are not so dramatic. For example, Collins has experienced two incidents of the same word being used for different ideas. Collins and Kusch, when writing their 1998 book, spent frustrating weeks in front of a whiteboard failing to comprehend that they were using the word 'action' in different ways, Kusch thinking of the term in a philosopher's way and Collins in a sociologist's way. For Kusch, an action was associated with responsibility in courts of law: if someone dropped a gun and it fired and killed someone, were they responsible for the action of firing? For Collins, this was not an action but an accident. For Collins an action was something that was formative of a society – for the Azande, consulting the poison oracle was an action but taking out a mortgage was not. In the UK, taking out a mortgage was an action but consulting the poison oracle was not. Only when they realized they were using the word 'action' in these very different ways, and it took a long time to realize it, could they make progress.

In the same way, for the sociologist Collins and psychologist Reber cats connoted very different things, though in this case they realized it almost immediately.[66] Below is a remark by Reber followed by a comment by Collins:

> *Reber:* [T]hink how you spent the morning ... you got up, scratched, rubbed your tummy, brushed your teeth, made coffee/tea (whatever), walked down the hallway [all unreflectively]. [...] It seems to me that you and the cat (and me) are rather similar ... (p. 140)

---

[65] From the point of view of the natives of the domains, this is collective tacit knowledge since no-one knows they have a tacit absence of rabbits and ducks respectively.

[66] Collins and Reber (2013).

Collins: [I see] the cat and the human as very different even when they are both doing things unreflectively. In particular, the cat cannot brush its teeth, make coffee/tea or even … walk down the 'hallway'—a hallway connotes a great deal to a human while a cat is just walking along an elongated space. Humans can only do things like brush teeth, make caffeinated beverages and walk in halls, however unreflectively they do them, because of the existence of a range of corresponding institutions linked together by language. Nothing the cat does is like this. Cats' activities are circumscribed by their evolutionary history; different groups of humans are, however, enormously different, the differences emerging from the reflective activities of other humans who are distant in time and space from what is going on now. … all these humans are linked together by a network of common social activity and language. So while both cat and human may be doing things in an unreflective way, most things that humans do depend on a history of reflection by other humans of a kind that the cat has not shared and cannot share since it has no language nor social life in the strong sense of social.

The point is, and this is doubly relevant to this book, the cat has not engaged in the face-to-face communication with humans necessary to root its understanding of hallways if that understanding is to match that of humans – another illustration of the 'Face-to-Face Principle' described in the Introduction. In short, cats do not have any interactional expertise.

Silence and sounds in Pisa

Here are two more examples of the duck-rabbit problem which can be found in Collins's 2004 book; these recount some experiences from his research on the sociology of the detection of gravitational waves. The first refers to the changing status of Joseph (Joe) Weber, the pioneer of gravitational wave detection who, in the late 1960 and early 1970s, was claiming to have detected the waves but who was largely disbelieved by the mid-1970s:

Thus, in conference after conference, Joe Weber would stand up and present his papers, explaining that he had found gravity waves long ago, and the delegates learned [from others' reactions] that the right response was to quietly move on to the next paper. And later, conferences would happen without the physical presence of Joe Weber or even his virtual presence in the vibrations of the airwaves that constitute words. In my first day at the [1996] Pisa [GW] conference, during which I listened to every paper, Weber's name was mentioned just once, in passing. (Collins 2004b, pp. 451–452)

There we see newcomers immersed in gravitational wave talk learning a piece of knowledge, without anyone having to say it. This piece of knowledge was of immense practical importance in the world of gravitational wave physics, namely that Joe Weber's early claims were wrong and any subsequent claims by Weber should be ignored. This information was so important that when Weber published a paper in 1996, claiming a positive result, it was impossible to find anyone from the community who had read it.[67]

The absence of sound representing Weber's name is, as has already been mentioned, only one kind of silence. Normal speech has a characteristic rhythm so that a long silence can sometimes be as meaningful as a word or phrase. For example, at a workshop a speaker might say something seen as inappropriate that is greeted with an extended silence equivalent to the spoken words 'that's an inappropriate remark'. The opposite is when a political speaker speaks across audience applause – where silence would normally be the response – actually inviting more applause.[68] Here when we talk of silence we are not talking about these meaningful interventions into the very rhythm of speech but the absence of certain words from the corpus; this absence helps to create the taken-for-granted-world of a society or a specialist domain without anyone necessarily having any idea that it is going on – just as in the duck-rabbit domains.

The second example is a positive rather than a negative one. Here we see words being used to create a new reality. The example is from the same, Pisa, conference; it is a contribution to the establishment of the reality of black holes – the possibility of the existence of which was, at that time, strongly doubted by a sizeable body of physicists. The quotation from Collins's 2004 book reads as follows:

> … at the Pisa conference, black holes were as comfortable and familiar as cups and saucers. The modalities surrounding the term *black hole* were those having to do with certainty. The theory of black holes was a matter of fact; this or that feature of black holes has not been postulated but 'discovered'. (Collins 2004b, p. 452)

Again, just by scientists speaking the way they spoke – and these were theorists talking, not observers, since no black hole has been directly observed even

---

[67] Collins (2004b, pp. 366 ff.).

[68] This distinction was brought out by Alun Preece. Later in the book we will indicate that there are three kinds of silence altogether, since the kind we are talking of here can be divided into two. It is Max Atkinson (1984) who first noticed the rhythm of politicians' speeches.

to this day – the ontology of black holes was being changed from theoretical speculation to discovered objects in the universe, and the epistemology of theorizing was being changed from a tool for hypothesis formation to a tool for making discoveries of real things. Of course, no-one said that this was what they were doing and, probably, the speakers only vaguely understood that this was what they were bringing about. But take away the generation of the spoken discourse at this and other face-to-face conferences, and the world or black holes would be different.

Gender and racial inequalities in the corpus

Feminists and others have long argued that social inequality is integral to the very language we speak.[69] It is increasingly possible to see the connections between language and social inequality by analysing massive collections of text. It is found that in English the words 'he' and 'him' will appear close to the word 'doctor' more often than the words 'she' and 'her'; through this statistical means we find that that the bath of language biases our thinking toward associating doctors with men not women. This is merely one instance of the way sexism is embedded in the relationship of words and silences in our language and there are obviously many others. Similar arguments can be made in respect of racism; the 'decolonizing the curriculum movement' in education, and campaigns highlighting the ways wealth generated through slavery is silently celebrated in the names of streets and buildings, all seek to highlight the ways in which public spaces and discourse reflect a very partial view of history. Among those who want to use corpus analysis techniques to train computers to be better at 'understanding' and transcribing language and the like, the powerful way analysis of bodies of words reveals cultural biases is a deep cause for concern precisely because training computers to understand cultures also transfers existing cultural biases in the population. Thus, Bolukbasi'spaper, aptly entitled 'Man is to Computer Programmer as Woman is to Homemaker', describes the gender biases in society that are revealed by analysis of the proximity of words and we know that other cultural biases have raised an ugly presence in other instances of artificial intelligence research.[70] Unfortunate though this is, it does

---

[69] There is a huge literature on gendered and racist language, and we do not attempt to summarise it here. For gendered language, classic texts include Lakoff (1973, 1975) and Butler (1990), while Speer and Stokoe's (2011) edited collection provides a good overview and introduction to academic research on the reality of everyday sexism in language. For racist language and practices, Allport (1954) remains a classic starting point, with Hill's (2008) *The Everyday Language of White Racism* providing a more recent analysis of the subtle racism found in middle-class American discourse.

[70] Swinger et al. 2019; Bolukbasi et al. 2016.

show the power of the method – the arrangement of words and silences reveals our cultural beliefs and biases.[71]

Why so little discussion of the practical significance of the corpus?

To show how the bath of words can bring about practical understanding, we have managed to dream up one just-so story, dredge up various examples from books written for a very different purpose, and refer to linguistic biases and the problems they cause for deep-learning computers! Why did we have to search so hard? The trouble is that language has mostly been thought of as something formal: it is thought to use commonly understood 'symbols', to explain things in an explicit way. But language is far more rich and subtle than that – our one or two examples are meant to show how it can work in this more subtle way.

It is not just words that are being learned in this kind of situation. At the same time, individuals learn what they can rely on – what they can take for granted in the society into which they are being socialized – they are making, through their existence and understandings, their contribution to the constitution of that society. They are, then, building the substance of *reliance* in that society. Reliance is learned along with language in the process of socialization. The bath of words is one of the things that teaches reliance, in part through silence – certain silences mean new members of a society are not warned against certain things so that certain other things come to be seen as reliable – something we will come back to. There is nothing much in addition to be said about it beyond what has been said about learning fluency in language; the transfer of reliance comes, of course, with acquiring interactional expertise and the fluency needed to learn from the bath of words happens in face-to-face communication.

Tacit knowledge-transfer and developing trust

As discussed earlier, part of what is acquired through socialization is not simply ways of speaking – it is also deeper tacit knowledge about the community. Much writing about tacit knowledge transfer is about practical skills. Collins showed in the early 1970s that, the inventor apart, those trying to build working 'Transversely Excited Atmospheric Pressure CO_2 lasers' (TEA-lasers) succeeded only if they visited the lab of someone who already had a working laser. Those who tried to build one, using even the most detailed published instructions, generally failed (always failed in the actual case study). Collins was later able to show some of the tacit features of the device that could, at the time, be transferred only by presence in the lab of someone successful. But, as has been explained here, developing fluency and developing reliance are also a matter of

---

[71] Investigate 'word embedding' via a search engine.

the transfer of tacit understandings. We now present a single case study which draws together the transfer of tacit knowledge and the development of trust among individuals.

Checkov

This case deals with the transfer, from Russia to the UK, of the tacit knowledge needed to measure a certain property of a crystal.[72] The 'Q' of an object is the time it takes for its vibrations to die away to half of their initial amplitude. If a bell is struck it will ring – the effect of vibrations in the metal – and they will slowly die away. If the metal of the bell is flawed, the vibrations will die more quickly. The purer the tone – the more limited to a narrow frequency band are the vibrations – the longer the bell will ring. In the hunt for the best materials for the mirrors of laser-interferometric detectors for gravitational waves, a material with a very high Q was needed so that the unavoidable ringing in the mirrors would be restricted to a narrow band that would not overlap with the frequency band of the gravitational wave signal and become confused with it. Any confusion of vibrations associated with the signal and vibrations in the mirrors would make the instrument less sensitive. The Russians claimed they had measured a very high Q for crystals of sapphire, but this measurement had to be repeated in the West if sapphire was to become a candidate for the mirrors of the Laser Interferometer Gravitational-Wave Observatory (LIGO). Note that the Q of sapphire is a difficult and dangerous piece of knowledge: Western scientists felt they had to repeat the measurements themselves before investing their time and effort into using sapphire in spite of the straightforwardness of the information offered by the Russians – 'sapphire has this very high Q, we have measured it to be XXX'. This bears on the difference of the importance between kinds of knowledge: if the measurement were invalid, investing in sapphire as the mirror substrate would waste millions of dollars. Still more important, it would probably mean the decades-long search for gravitational waves – something in which the scientists were investing their lives – would end in failure. But try as they might, Western laboratories could not make sapphire crystals ring for anything like as long as the Russians said they could make them ring.

Eventually, a Russian scientist, to whom we give the pseudonym 'Checkov', travelled to a laboratory in Glasgow, bringing with him some samples of sapphire, to show the British scientists how to accomplish the measurements. The key is the way the crystal is mounted in the benchtop apparatus so as to avoid leakage of energy from crystal to its support or to the remaining air in the evacuated container (the details can be found in Collins's 2001 paper). To master the art requires many repeated trials with slightly changed conditions

---

[72] This section is based on Collins's (2001) paper 'Tacit Knowledge, Trust and the Q of Sapphire'.

each time. It takes a long time to pump down the vacuum in the apparatus every time the conditions are adjusted. The Glasgow team led by Checkov worked away, running trial after trial for several days but without ever achieving the high Q claimed by the Russians. Eventually Checkov had to depart but left a crystal with the Glasgow group that he assured them was pure enough to yield the appropriate Q if they stuck with the procedures he had taught them for long enough. The Glasgow group persevered but could not accomplish the measurement. Then Checkov told them the specimen must be flawed after all and not the perfect piece of sapphire he had said it was!

This seems a classic case of scientific fraud. But the Glasgow group still decided to press on. Collins asked them why they would do this when everything was telling them that the Russian claim must be false. The answer was that in the course of the few days of joint work they had come to have complete trust in Checkov and were inclined to believe everything he said. Collins asked them why. They replied in terms of the way Checkov worked at the bench – it was convincing if you were there. Here is a typical exchange (Donald is a pseudonym):

> *Collins:* Let's push this: can you really tell me how you came to this conclusion?
>
> *Donald:* Well, sitting in front of this apparatus to a large extent – him looking at what we were doing and he would say 'I want to try something and modify something slightly' and you'd see improvements taking place. And he would say if you changed something you'd make it worse, and, right enough, you would change it and it would get worse. And also, you know, you hardly needed to exchange words – it was one of these things. You were thinking the same way and that is how we made such enormous progress. Because the interactions were very good with the man – you could tell how he was thinking and he could understand how you were thinking.
>
> *Collins:* And there was no way this could have happened unless he'd actually been here, or you'd been there.
>
> *Donald:* No – you need to have someone actually working in the lab; we were just gathered round this machine. This summer when he was across, we spent 90 hours in the lab from starting on a Sunday and finishing on the following Sunday. And he didn't want to go out and eat. He much preferred just to quickly get a sandwich and come back, and just keep going, and so we worked like that for seven days, and it is very impressive when you have a small group working like that. You get a lot done.

The Glasgow team persisted and eventually they did achieve the promised measurements but their persistence – and remember this is a long-drawn-out and exhausting business – was a result of their complete trust in Checkov;

without that, though it is a counterfactual claim, they would almost certainly have given up. So one can see that trust is an important part of the transfer of tacit knowledge – without trust there would not be the persistence required to learn the skills through repeated trials in a situation where it might be more reasonable to think that those skills do not exist because what they can achieve has been misdescribed: imagine, would one have the persistence to learn to ride a bike if one did not know that it could be ridden?

Now consider the element in that vital trust associated with the fact that Checkov did not want to go out to eat when he could have enjoyed delicious meals in Glasgow's excellent restaurants, turning his trip from Russia into at least, a partial holiday. He preferred to stay in the laboratory working, and this helped convince the Glasgow scientists that this man was serious. We'll call this 'The Checkov sandwich'.

What would remote means of communication, even with as high a bandwidth as trimensional viewing, or even virtual reality, have to achieve to generate trust in the way of the Checkov sandwich? It would have to achieve the illusion that Checkov had actually made the arduous journey from Russia and was present in Glasgow, potentially able to enjoy the food for which it is renowned, such as world-class curry. That refusal to leave the lab and eat in restaurants would have to be as palpable as was the case with Checkov's actual co-presence. Can we say this level of virtual reality will never be achieved – after all we can imagine Checkov in his turn being able to enjoy the sensation of virtual Glasgow meals if only he was willing to take time away from what would be his Moscow-based bench? The answer is that we cannot say this will never happen, but setting aside the necessary deceit in terms of Checkov's travel from Russia to Glasgow, which could negate the whole trust-building exercise, we can say it is not likely to happen in a world that is ever going to be of interest to us – it tells us where the virtual reality makers will have to go if that element of face to face is to be mimicked.[73]

---

[73] There is a scene in the film *The Matrix*, which concerns the supposed invisibility of the borderline between reality and virtual reality, where Agent Smith, a machine, and the human, Cipher, are eating in a restaurant when Cipher betrays Neo and Morpheus, his human friends, to the machines. It might be interesting – in terms of this discussion of eating as engendering trust – that the restaurant is used as a place of betrayal to the machines who do not eat (Agent Smith does not eat during the scene). Also, Cipher talks about knowing the food he is eating is not real, but that it is still delicious – so the Matrix is made out to have created the conditions for him to have bodily experience that are better than the real world (the steak he eats is much nicer than the protein-gloop he gets in the real world). But he also states that he wants to forget that what he is eating is not real – he does not want to know that he is in the Matrix, as this knowledge detracts from the value of his experience. If one is conscious of the fact that one's

Eating and drinking together – commensality – is going to come up over and over again in this analysis. There are two elements to it – the first is that sharing food or, at least, eating together, engenders trust. Perhaps this emerges from something deep in evolution or psychology: animals share food only with their families or their immediate group often having to fight off scavengers to keep their share; drinking at a waterhole is a time of danger and being willing to drink alongside another animal indicates that no attack is anticipated. In a modern life of conference-going there is plenty of food and drink and danger is *not* ever-present, but perhaps something from our past has survived. The second element, more immediately relevant to modern life, is the way food is consumed – the style of eating and drinking. That Checkov chose a sandwich is telling us something about Checkov's priorities, something difficult to reproduce without co-locality. And who pays is a powerful indicator of colleague-ship: buying someone a drink, inviting them to dinner – maybe at one's house – all these things tell the other that one is committed to a relationship. Or when the group at a conference goes out for a restaurant meal and works out how to split the bill, quite a lot is going on. Which of the group will eat a starter, the most expensive dish, and a dessert and drink the wine and maybe a liqueur and then say, with a generous-sounding flourish, 'Let's just split the bill evenly'; which of the group is going to insist they pay for what they consumed, perhaps to save their own money or perhaps so as not to exploit others? One quickly learns who to watch out for in any future collaboration.

Classifying tacit knowledge

Tacit knowledge is of central importance in understanding why face-to-face communication cannot be replaced so it is worth spending a little time on it. The term 'tacit knowledge' is usually attributed to Michael Polanyi, but as mentioned in the introduction, Polanyi's vision was much concerned with individual intuition whereas, here, what we are discussing has to do with kind of knowledge that either can or cannot be transferred *between people*, using different kinds of communication. Notice that in discussions of tacit knowledge there is often confusion about whether the tacit is something that is not known explicitly and thus cannot be transferred by explicit means of communication, or is something that *cannot be* known explicitly and *cannot* be transferred by explicit communication. A three-way classification of tacit knowledge resolves many of the confusions.[74]

---

reality is virtual then even sharing a delicious 'matrix' meal together would not increase trust in the same way that the Chekhov sandwich does because the effort/personal sacrifice required is less.

[74] Collins (2010).

There are three basic reasons why knowledge either *is not* or *cannot be* transferred explicitly. The first reason is that knowledge that could be transferred in principle by explicit means is not transferred because, for example, the parties do not know it needs to be transferred or the ways of explicating it have not yet been developed. An iconic example is the need for the length of the top lead of an early TEA-laser to be short so as to minimise inductance. There was a time when successful laser builders, copying the design from others, would make the lead short by mounting the heavy capacitor connected to it upside down in an elaborate frame as others had done, whereas those who worked from circuit diagrams would naturally mount the capacitor in the bench, meaning their lead would be too long and the laser would not work. There was no need for any of the parties, either learning or not learning from each other, to understand what was going on. Later, the role of inductance in the top lead became understood and the knowledge became explicit. The first category, *relational tacit knowledge*, or RTK, is knowledge that is not transferred for reasons of this kind – reasons that could be resolved in principle but are not. In practice there is far too much RTK in world for it ever to become explicated all at once and as the frontiers of science and technology and other practical skills move forward there is always a window of activity where tacit knowledge of this kind will feature in the transfer of skills and techniques. One can see RTK at work in the Checkov example: for instance, a lot of the techniques of experimenting fast and efficiently could have been explained but no-one knows how much of it needs to be explained to transfer the knowledge successfully – here demonstrations resolved the problem.

The second type of tacit knowledge is '*somatic tacit knowledge*' (STK), which is the 'knowledge' contained in the very substance of brain, nerves and muscles that enable humans to execute various skills, sport being the example most often referred to. In scientific work, the possession of these kinds of skills, combined with relational tacit knowledge, can make the difference between the 'golden handed' experimenter, for whom every experiment works, and the rest, for whom almost nothing works. Somatic tacit knowledge is, again, explicable *in principle*, as the advance of robotics demonstrates, but the very slowness of the advance shows how difficult it is. Strangely, what were once thought of the frontiers of human intelligence – mathematical abilities – have fallen rapidly into the domain of computing while practical abilities continue to resist. As with relational tacit knowledge, the resistance is not a matter of principle in respect of any specific element of ability or knowledge but a matter of resources and the impossibility of explicating every part of the domain at once.

The third kind of tacit knowledge is more resistant to explication and is the strongest candidate for being *impossible* to explicate. This is '*collective tacit knowledge*' (CTK). The iconic example here is natural language. The reason natural language processing is so hard to reproduce mechanically is that a natural language is not the possession of individual but the collective – the society that embodies it. Natural languages are continually changing as society changes, and it happens in unpredictable ways. Therefore, for a machine to be as fluent in natural language processing as a human requires it to be embedded within the human collective in the same way as a human and, to date, we do not know how to do this. The invention of deep learning has marked a huge increase in the ability of computers to handle natural language precisely because self-teaching devices are able to access the internet, which is a superficial and narrow reflection of society. This, however, is still far from full embedding in society though it is much better than anything we have had before. Its shallowness is easily revealed when even the very best of modern deep learning machines are exposed to demanding Turing Tests, including, for example, the Winograd schemas described above, and some of the other tests of fluency already covered.[75] Other examples of collective tacit knowledge are negotiating traffic – the skills of which vary hugely from country to country – or, indeed, any of those things that are referred to as culture and appear in the top oval of the fractal model of society. This classification of tacit knowledge should help to demystify it while, at the same time, explaining why local interaction is so vital to its transfer. It also shows why language learning can stand in for different kinds of skills learning when we consider whether face-to-face interaction is important; language is vital in the establishment of legitimacy (are they ducks or rabbits?) and as the medium of interactional expertise, and natural language is also an excellent example of tacit knowledge acquisition.

Domain discrimination

Face-to-face communication also affords the development of 'domain-specific discrimination' or 'domain discrimination' for short, which once more emerges from studies of science. From the outside it may appear that science is a formal process of hypothesis and test, but closer examination shows that this is only the visible carapace of a dense network of trusting relations. Trusting relations are vital because, as already intimated, if a scientist is going to act – which means investing time and resources – on the claims of some other scientist, they have to know how much trust to put into the claims – illustrated by Checkov and the Q of sapphire. The trouble is that scientists cannot test those claims according to a formal procedure because experiments rest on tacit knowledge and tacit knowledge is, as we have seen, hard to acquire. This is just another face of the intimate network of relations between trust and tacit knowledge. We can illustrate it further with the remarks of scientists.

In September 2015, just a couple of weeks before the momentous first discovery, Collins had a series of recorded discussions at a gravitational wave conference in Budapest, asking scientists to explain why, given that they were so

---

[75] These are explained in Collins (2010, Ch. 10).

well connected over the internet, they still travelled across the world to attend conferences and workshops.[76]

By the way, this book aspires to an interdisciplinary audience and some of our readers might expect much more in the way of scientific procedures – surveys with mathematical analysis supported by significance tests and the like. But not all sound social research has to be like this. For example, suppose I want to discover the position of the verb in the English sentence. A few sentences spoken by a haphazardly selected handful of volunteers would be sufficient because the phenomenon is 'uniform' and where a phenomenon is sufficiently uniform even a single 'token' can be representative of entire society. (To see the logic, ask why physicists were not worried about whether Swiss Higgs bosons represented all Higgs bosons.) Ethnographic, anthropological and sociological research which has space for the notion of the 'native informant' uses this logic, albeit implicitly. In this book we are largely dealing with the uniform properties of societies.[77]

Going back to 'the Budapest meeting', one scientist remarked:

> Part of being an expert is understanding whether what's being said is legitimate, what result people believe and what they don't believe and what's out of fashion, and you can't get that unless you are around other people and you're seeing their reactions. You're sitting at a table at lunch and you bring out some paper or some theory and you can see how everyone reacts. And if you're writing letters to them asking them [that won't happen] …

To return to trustworthiness as initially discussed above under the heading of 'domain discrimination', as a fieldworker Collins found that gaining a sense of who was counted as trustworthy and who was not was far harder and required far more continuous immersion in the group than even acquiring the specialist technical tacit knowledge of the group; technical tacit knowledge tends to be relatively enduring whereas knowledge of who is trustworthy changes every time someone new joins the group.[78]

In science, trustworthiness is essential since, as we now know, an experimental result, even a peer-reviewed and published experimental result does not speak for itself. To know which results to take seriously scientists have to know which other scientists to take seriously and this assessment takes place in personal interaction and in the patterns of words and silences found in small

---

[76] The results are reported in full in Collins (2018, Ch. 8).

[77] For a full analysis see Collins and Evans's (2017a) paper entitled 'Probes, surveys and the ontology of the social', a shorter version of which can be found in Collins (2019, Ch. 9).

[78] This is explained in Collins (2017, pp. 321 ff).

group conversations. Because scientists know that experimental results are always disputable, it can be that those far from the frontiers of science are more impressed – either positively or negatively – by experimental and observational results than are the scientists themselves. This phenomenon has been summed up as 'distance lends enchantment', and it explains quite a bit of the interaction between scientists and the public, leading commentators to imagine, too easily, that local trust relations can be replaced by technically mediated remote interactions. It means, inter alia, that if one's sole source of scientific knowledge were the scientific journals one would simply not understand science, since a large proportion of what is found there is wrong or unreliable and the only way to sort it out is to through the process of specialist scientific socialization in local groups.

Scientists, then, cannot make progress in frontier fields such as gravitational wave physics unless they know who to trust and who not to trust – they have to discriminate between those who are offering claims and information in much the same way as we have to discriminate in ordinary life. The process of becoming socialized into a specialist scientific field includes picking up on how much credibility to bestow on the remarks of various identifiable persons – indeed, this is one of the most demanding parts of being socialized into one of these domains. Toward the end of his study of the field, when he was travelling to fewer conferences, Collins found his understanding of the technicalities was far less damaged by the decrease in density of contact than was his understanding of the people who were entering the field and he had to rely on others for this aspect of his knowledge. Here are a couple of indicative remarks about different problems, one negative, one positive, sent to Collins by email from friendly scientists at the time of the final triumphant discovery process [pseudonyms]:

A very weird argument indeed from [Podolsky], although perhaps not so weird considering it was from [Podolsky];-)

I think before we're done we are going to have to understand whether there is any credibility to that … and I think that's going to be a struggle because [Quaglino] is a really smart guy and he's pretty self-confident and he will say he believes it and people have enough respect for him that they will not blow him off so I don't know how we're going to resolve that.

Summary: socialization, reliance and trust

All these detailed matters of trust take place against a background of reliance. In a science such as detection of gravitational waves, the frontiers involve trust such as that which led to the reproduction of the Russian claims but as time goes by these become so thoroughly accepted that those socialized into the field come to rely on them without thinking about it. The quantitative result for the Q of sapphire will now be relied on. When the measurements were being made in Glasgow, the scientists as they learned to trust Checkov, were, of course, relying on the outcomes of countless earlier such incidents which led to their reliance on clocks, interferometers, vacuum pumps, the properties of electricity, and so on and so on. The transactions that led to trust and then on to reliance for some of these historical incidents are what have been studied by historians of science with an interest in the sociology of knowledge.

So far, we have moved from primary socialization and the establishment of early reliance and linguistic fluency, including what is latent in the distribution of sounds and silences, through ubiquitous reliance and ubiquitous practices, to trust in individuals and the need for trust in the development of new ideas, which become part of the unspoken background of specialities and societies as they evolve. In the next chapter we will complete the exposition of Table 1.1.

CHAPTER 3

Completing the Story of Face-to-Face Communication

We now move on to the third and fourth groups of Table 1.1, that is the second six features of face-to-face communication as we have listed them. To start with, an important feature of face-to-face communication is that in the to-and-fro of discussion, immediate and reactive attempts can be made to clarify meaning. This decreases the possibility that things will go wrong. In Plato's *Phaedrus*, Socrates says:

> ... it shows great folly ... to suppose that one can transmit or acquire clear and certain knowledge of an art through the medium of writing, or that written words can do more than remind the reader of what he already knows on any given subject. ... The fact is, Phaedrus, that writing involves a similar disadvantage to painting. The productions of painting look like living beings, but if you ask questions they maintain a solemn silence. The same holds true of written words; you might suppose that they understand what they are saying, but if you ask them what they mean by anything they simply return the same answer over and over again. Besides, once a thing is committed to writing it circulates equally among those who understand the subject and those who have no business with it; a writing cannot distinguish between suitable and unsuitable readers. And if it is ill-treated or unfairly abused it always needs its parent to come to its rescue; it is quite incapable of defending or helping itself.[79]

The spoken word is not quite so defenceless as the written word and is therefore a bit better at carrying meaning. Phaedrus's problem is brought out by

---

[79] Hamilton (1973, p. 275), which is also quoted in Collins (1990, p. 11–12).

How to cite this book chapter:
Collins, H., et al. 2022. *The Face-to-Face Principle: Science, Trust, Democracy and the Internet*. Pp. 67–78. Cardiff: Cardiff University Press. DOI: https://doi.org/10.18573 /book7.d. Licence: CC-BY-NC-ND 4.0

French literary criticism's insistence that the meaning of texts is continually created anew by its readers, the opposite of what we intend in scientific communication.[80]

Presence, body language meaning and disagreement

It is convenient to deal with the second and third items belonging to the third main category of Table 1.1 – items 8 and 9 – together because the points are mixed up in the illustrative material. These two items are to do with the body language of co-present conversationalists, which creates and modifies meaning and enables disagreement that would almost certainly cause rupture in remotely conducted relationships to pass without doing too much damage, or even to be productive.

We have mentioned the way that the physical presence of mothers and fathers reinforces early language learning, with touches and cuddles along with intonation, and the direction of gaze, showing the infant when something is important or less important, and where something starts and where it ends. Likewise, the infant can demand repetition and reinforcement via physical cues. Collins's two-year-old granddaughter used to run to him when they first met after a long absence signifying a huge amount that is not expressed – there is no running to someone in remote communication. The bath of language will not work properly to bring about primary socialization without these physical cues.[81] Something similar applies to the touches, nods, winks and physical states that are readable only to someone close by. Influential studies were conducted by social psychologist Albert Mehrabian. He is often reported, mistakenly, as claiming that 7% of meaning is in the words that are spoken, 38% of meaning is paralinguistic (the way that the words are said) and 55% of meaning is in facial expression.[82] His claims were actually about the extent to which people came to like or dislike others as a result of these features of communication, not about the transmission of meaning. Whatever, the claims will certainly bear on the development of trust, and Mehrabian has been influential in drawing attention to the features of face-to-face conversation. As he wrote to Collins:

---

[80] E.g. Barthes 1968.

[81] In Collins (2018), the author provides an example where he completely misread the body language of an important claim that was being made to him and totally misinterpreted it: a senior scientist told him that no-one could believe X but what he was saying was not that no-one believed it – Collins soon discovered it was the majority view – but that the view was surely mistaken.

[82] See e.g. Mehrabian (1972). For discussion of the misinterpretation of Mehrabian see Wikipedia (2021a).

Just picture an angry individual in front of you vs. the same angry individual on a video screen. There is a world of difference between the two experiences with arousal level considerably higher in the former situation. (private communication to Collins, 3 October 2018)

Elaborating the point, with co-location one might detect blood-flow patterns on the skin or sweat indicating arousal or tension along with changed speed of breathing.

Quotations from the Budapest meeting (first discussed in Chapter 2, Domain discrimination) once more capture the way co-location works for physicists. Pseudonyms are used:

> *Delta:* When the collaborators are in different places you can have tensions and misunderstandings can build up. And often you get really frustrated with someone and I think they are doing things wrong and they're misunderstanding things and maybe we're just not going to continue working together because this is not going well.[83] And then you see them in person and you have a beer and you chat about things and then it's all fine. If you don't have that every so often, things can get very complicated – for no good reason – just misunderstanding – paranoia … Maybe they're trying to get to a result before you, or they've got some hidden agenda. To some extent we're all trying to do this, we're all trying to get ahead, but the fact that you are working together and you respect each other and you want to continue working together – that doesn't come across in a telecon[ference] or certainly not in emails … I think people who don't go to meetings and stay away – I think they get wound up in their own world.

---

[83] A quite startling instance of this occurred in this project itself in early January 2019, when one of Collins's emails was interpreted as an insult by one of the project participants. Fortunately, the participant re-interpreted before the following day and offered an apology before Collins had worked out how to respond. Baym (2015), Ch. 3 discusses 'flaming' – a feature of remote communication. Flaming is the self-conscious hyperbolic use of strings of insults in the course of a remote argument. But here we are talking about something different though it has the same outcome; here we are talking about sharp exchanges using remote media that unintentionally escalate into damaging and often irreparable dialogue because it is not moderated by body language, other kinds of social interaction such as commensality, or the mutually understood need for self-control which is present in F2F. It is not unusual to hear tales of the unintended rupturing of friendships as a result of the adverse effect of accidental flaming or accidental interpretation of relatively mild disagreements *as* flaming.

The reference to 'having a beer' in this remark echoes the importance of eating and drinking together – commensality – that was first discussed under the 'Checkov' heading.

> *Epsilon:* Often this is a great way to trigger really new thought and not just walk along saying the same things that everyone is saying, just in a different way. Sometimes you want to say things that are pushing it a bit harder just to see how other people react to it. And that's how a new thought emerges. You don't want to come across as being a provocative type and too stupid to understand the common consensus, but you just generally want to try out things.

This remark bears on item number 8 in Table 1.1 and other comments that have been made in the text: it is much easier to disagree without rupturing a relationship in a face-to-face situation because body language shows that the business at hand is not making enemies but mutual productive exploration of a disagreement, as expressed in the following remarks.

> *Epsilon:* I think sometimes it's useful to have this role-play where you say something and I deliberately try to find arguments against it, though my instinctive reaction might be to agree with you but doesn't help, so I try to be critical and I say 'No – why isn't this and this?' Some of the very best discussions I've had – some of the best interactions and experiences in science are in a small group of people when you can say 'No that's wrong' and you can argue about it, and there is a sense of trust, the person doesn't think you're an idiot: they know you're smart and you're confident and you're asking a real question, and they're honestly trying to explain it to you and you're honestly trying to understand it, and from that ideas flow and one of the most interesting results, the YYYY thing, came from a week of – there were a few of us – and at the beginning of the week someone says, 'I think the thing looks like this,' and then someone says, 'I think it looks like this, no it's not like that,' and back and forth and arguing, and within a week we'd found something new that we truly didn't expect at the beginning.

Commensality promotes trust and helps dissolve misunderstandings whereas remote communication can easily amplify misunderstandings; knowledge is often acquired and generated in circumstances where trust allows conversations to be adventurous or provocative in a way that would be too risky in remote interchange. It might be worth noting that these Budapest responses were themselves generated in an ambience of commensality – over lunch.

The establishment of meaning as well as the triggering of new thoughts also happens locally in a way that was described by *Delta*, continuing his remark above:

> Part of being an expert is understanding whether what's being said is legitimate, what result people believe and what they don't believe and what's out of fashion, and you can't get that unless you are around other people and you're seeing their reactions.

Bodily co-presence

Trustworthiness is established through talk but also through bodily presence – who wants to sit near who – who wants to talk to who. As with commensality, there are direct and indirect faces of co-location. The direct face has to do with simply spending time in someone's presence, which represents a cost to both parties and shows that each considers the other worthy of spending time with and worth conversing with. The way the interaction goes is another generator of trust (or distrust): 'I' can give 'you' a certain look, or a certain grimace when I say a particular word, and I can shake hands, or reassuringly touch your upper arm as a gesture of friendship, or react to a positive remark of yours with something subtly less than enthusiasm so as not to be rude enough to criticise what you are saying but to indicate in the most un-confrontational of ways, that I do not share the enthusiasm, and so on – in sum, it is a matter of 'body language'. The indirect face has to do with others observing 'you' interacting with someone else, preferably someone important, showing them that 'you' are trusted by important people and can, therefore, can be trusted by them too. Thus, in a small group, the way people listen to others – this person or that person – or the way others react to a remark by a third party, can say a lot about how credible the person or the remark is to be taken to be. The group continually creates meaning by the way it attends. On top of this there will be more formal affirmation or contestation – nods, shakes of the head or straightforward agreements and disagreements, but reinforced in group reactions. Quotations from the Budapest interviews have already pointed to some of these features.

Collins, himself, gained hugely in terms of trust from eating at the 'high table' of gravitational wave meetings on frequent occasions, the whole process being self-reinforcing – trust engendering more indications of trust and so forth. The following is Collins's account of the very start of the process of being accepted into the 'big science' gravitational wave group, the date being 1996:

> 'The International Conference on Gravitational Waves: Sources and Detectors' was held near Pisa between March 19 and 23, 1996. Like most of the attendees, I flew in the day before, and spent a very pleasant

few hours in the first sunshine I had seen in several months, wandering around the city and inspecting the Leaning Tower...

Pisa was chosen as the site for this conference because VIRGO, the European laser interferometer, was to be built just a few kilometres away. The organizers had arranged for buses to take the delegates from their Pisa hotels to Cascina, the small town where the meeting would be held and where VIRGO would be built. From about 8.20 to 8.35 the following morning, small groups of men and a few women could be seen standing around in front of Pisa railway station, wearing that vacant look that comes from being in an awkward social situation. We were probably but not absolutely certainly waiting for the same bus to take us to the conference. Were we supposed to know each other or not? To avoid embarrassment, we had to act like any other set of strangers, no-one catching anyone else's eye.

The buses pulled up, and as each person embarked on a manifestly individual trek to one or the other he or she passed through a literal as well as metaphorical door, becoming one of a group of colleagues going to a conference. Still feeling pretty lonely, I boarded, too, and was delighted when my [physicist] pals from Frascati stepped onto the same bus and, recognizing me, shook my hand and exchanged some pleasantries as they passed down the aisle. I had now gained a little status that I could use in my work; I was somebody, not nobody—somebody that physicists spoke to and someone whose hand was worth shaking. ... I felt sorry for the [few] lonely physicists whose manifest isolation spoke volumes.[84]

A reflexive look at face to face

We have some empirical material from our own experience on how turn-taking is organised using remote media on the one hand and face to face on the other. This is to do with the discussions that that gave rise to the writing of this book. The project – originally a grant application (it was not funded) – was set up using emails. This, as already explained, was possible because the participants had already developed a degree of trust and understanding through personal contacts – they already knew each other. Very roughly, something in the order of 1,000 emails were involved in setting up the project and its associated grant application. When we learned our application was not successful, Collins immediately wrote to all participants and asked if they were willing to go ahead without funding given the interest of the project; all agreed (there would be a

---

[84] *Gravity's Shadow* (Collins 2004b, pp. 449–450).

Table 3.1: Turn-taking in first 1hr 47mins of the F2F meeting.

Times/Interaction Types	Affirmations (As)	Disaffirmations (Ds)	Proper Turns (PTs)
Hour 1	272	22	352
Next 47 Minutes	170	7	173
Totals	442	29	525

few drop-outs but this was almost certainly nothing to do with the lack of funding – it is the sort of thing that happens with already overworked people with their own projects, sometimes in very different fields).

A crucial feature of the grant was to be a face-to-face meeting of our own to take place within a month or two of the start of the project. Participants agreed that this would take place even without the grant, with lunches, teas and coffees covered by Cardiff School of Social Sciences, while two attendees from North America paid their own fares and hotel bills – and thus concretely demonstrated commitment to the project, inserting energy into it. To continue to arrange the project in its new form, up to the moment of the start of the F2F meeting – the morning of 8 March 2019 – was again a matter of large numbers of emails.

We decided to take a reflexive look at the face-to-face meeting and compare what happened there with what had been achieved in the email interchanges. Therefore, the first hour or so of the meeting was recorded and subsequently analysed. We found that in the first hour and 47 minutes there were a total of 961 conversational turns. Of these, 525 were substantive turns involving a change of speaker, which either developed an existing theme or changed the theme. And 442 were affirmatory interjections, such as 'yeah', 'uh-huh' and 'right', that did not lead to a change of speaker, while 29 were similar but negative in intent such as 'don't know', 'well' and 'hmm'.[85] Table 3.1 analyses the data for the first hour and the subsequent 47 minutes.

The face-to-face meeting continued over two days for about 16 hours in total with most of the conversation being equally intense. If we make a very conservative estimate, it would have involved in the region of 8,000 conversational turns, which we would also estimate would take about a year if conducted by email among seven people.[86]

But much more was achieved at this meeting. This part of the reflexive look at our own work refers back to the first item in Table 1.1, the way commitment is

---

[85] Will Mason-Wilkes was responsible for the recording and analysis.

[86] For comparison, the email discussion over five months among the 1,000-plus gravitational wave physicists responsible for the confirmation of the first discovery of a gravitational wave involved 12,000 emails.

expressed by the effort needed to co-locate. As already intimated, the commitment expressed by the two co-applicants who travelled from North America
was an enormous boost to the project and the fact that the other co-applicants
were prepared to give up two days of their lives, including a Saturday, and engage
animatedly for the entire period, turned the project from an idea into a reality.
Of the seven participants, only one, or perhaps two, could be said to be part
of an existing team – Collins and Evans have many co-authored projects and
they jointly supervised Wilkes's PhD so you could say these three were bound
by a common definition of what would count as successful academic work in
their lives. While the others knew each other to a greater or lesser degree, they
were committed, mostly extremely heavily, to different kinds of academic work;
their being initially drawn into the project depended on their seeing that the
question of face-to-face interaction was an interesting and important one and
that they would be willing to be co-applicants on the grant that Collins had put
together in the first place. But, to maintain that commitment after the rejection
of the grant was another matter. The success of the first face-to-face meeting
was the mortar that stabilised the continuation of the group.

The meeting also delivered a lot in terms of substantive understanding,
with the book developing a number of new themes emerging out of quite
marked confrontations that would have been hard to manage over email; these
began with sharp and profound criticism of some of the assumptions that
had been brought into the meeting from the beginning. At one point, Collins
announced, in the face of a severe criticism: 'I am stumped'. At that moment the
whole project seemed to be slipping away. But shortly afterward a way forward
emerged from the interplay of discussion. It could be that this would have happened over email, but it would not be something one would bet on.

For now, let us note that face-to-face conversation allows turn-taking at
breathtaking speed, not only between two individuals but also in a small gathering of speakers. A rapidly flowing series of conversational interchanges can
be instantly organised by a co-located group via the bodily interaction, and
'metadiscourse' – talk about what the talk is doing (a kind of elaborate set
of spoken emoticons).[87] In face-to-face groups, when they are working well,
people get their turn and take their turn mostly without strain and with far
less chance of misunderstanding than happens remotely. Body language plays
a big part in this, moderated by the mutual understanding of status relationships and, of course, intonation, with group pressure being a well-documented
determinant of behaviour in such groups. As already remarked, proximity also
allows the kind of condensed and emotive contribution to a conversation that
would be considered rude in a remote setting. As scientists at the Budapest
meeting indicated, in face-to-face interchanges, questions and responses were
exchanged 'at the speed of thought', so that a long sequence of exchanges could
be completed in a very short time and with a number of people. Face to face

---

[87] See Mauranen (2003) for metadiscourse.

allows interjections and the switch of a path of a conversation without planning, sometimes with the parties not understanding why the wind of discussion blows one way rather than another. Trying to accomplish this via, say, email with multiple persons would be impossible, especially if different people have different email-using conventions and expectations.

Moving to the second line of the final section, bodily co-presence allows for many meetings with many people to take place either through arrangement or spontaneously. As one of the Budapest meeting scientists said to Collins:

> *Delta:* Here [at a meeting like this] you can just go round a bunch of people and in ten minutes the whole thing [a new collaboration] is arranged and you can tell immediately whether the people are enthusiastic, how interested they are, how much commitment you are going to get from them just from their reaction, which you would have no idea about over email. In email [you get]: 'That's great! I'd love to do that' [but it does not mean anything.]

Then again, one can bump into people entirely fortuitously in the corridors and coffee bars and have unanticipated interactions, and sometimes these serendipitous encounters are of enormous value. Again, quoting a Budapest meeting scientist:

> *Alpha:* I can certainly recall occasions when random conversations over lunch and coffee have triggered some insights or ideas. Chance conversations can generate these things.

Small face-to-face groups and science

There seems to be something truly universal about trust in small, local, groups. From babyhood we learn to trust and rely totally on our immediate family, and we learn to trust our 'tribe'. In modern societies we learn from childhood to distrust strangers as in older societies we learned that other tribes were our enemies. Network analysts stress the importance of 'homophily' – love of those who are like us; we are more likely to learn from those like us.

But, as we mentioned in the Introduction, there is a caveat: small, face-to-face groups do not have to be directed at benign ends. Small groups are efficient and trust-inducing even where the purpose is something quite different. Positive uses might include the formation of faith communities and charity groups or the use of personal recommendations for tradespeople amongst friend and neighbours. Negative uses include the importance of face-to-face interactions in illegal activities ranging from organizing a confidence trick, a bank robbery, a terrorist cell, to the mass suicide of a cult. Face-to-face interaction creates opportunities for the use of verbal and non-verbal cues, including inauthentic

warmth, confidence and sympathy, to convince people to make bad investments, fail to vaccinate their children, and so on. For some, personal interaction can be a trap, cutting off the opportunity for reasoned reflection. This
has been experimentally demonstrated by the way local group pressure can
persuade people to mis-classify the length of lines drawn on paper (as in the
famous Solomon Asch experiments published in 1951).[88] Furthermore, the
Nuremberg rallies were examples of face-to-face communication if not in small
groups, and so are such contemporary political rallies that seem to echo them
in style if not in substance.

Could it be that there are special qualities that separate small groups into
types and that our use of science as an example is more than a contingency of
the professional careers of one or two of us? Science is directed at truth-finding
and, as already remarked, is largely driven by intrinsic goals rather than desire
for power or riches. An old theme in the sociology of science is that this is
associated with certain norms.[89] But face-to-face groups vary on a number of
dimensions. Four that seem to be useful in understanding and demarcating
science as a social institution are:

1) Are the groups large and public or small and intimate?
2) Are the groups open or closed: who is allowed into an intimate face-to-face
    setting and how is the boundary maintained?
3) Is the organization of the group flat or hierarchical?
4) How much dissent does the group allow?[90]

If we take science as an iconic example of where things mostly work well, we
can, perhaps, learn from it about small groups in general by describing it on
these four dimensions. In science the crucial face-to-face interactions are small
and intimate – like a family with the same kind of trust and disagreements
you get in a family. These small intimate groups are called 'core sets'.[91] In 'big
sciences' the core set can be a thousand or more strong but still characterized

---

[88] Our colleague Mike Gorman points out that sometimes a single voice can
dominate a local interaction whereas a slow medium like email can offer
time for reflection.

[89] Robert Merton (1942) set out five norms, including two we will discuss here
– 'organised scepticism' and 'universalism' – as the constituting features of
science. Collins and Evans (2017b) extend the list of values of science from
5 to 14, including certain standard supposed elements of the 'logic of scientific discovery' (Popper 1959), such as corroboration and falsification
which Collins and Evans treat, not as part of the 'logic of science' but as
constitutive values of science which characterise the institution, not every
individual action.

[90] We will take another look at classifying groups in Table 11.1.

[91] The idea of the core set is fully set out in Collins (1985/92). Core sets can
have schisms where there are deep disagreements.

by small-family levels of trust, like a tribe that maintains styles of interaction typical of small groups.[92]

The small groups in science erect strong boundaries around themselves. Thus, to become trusted by the gravitational wave core set, Collins had to work hard. In the normal way no-one would be allowed to attend one of their conferences or workshops unless they were an established member of the scientific 'collaboration' which involved signing a 'Memorandum of Understanding' (MOU) detailing the *contribution* to be made to the collective research project and involving a formal presentation vetted by the community; Collins was not offering an MOU, nor promising any other contribution to the field, but rather, preserving his independence, and was the only outsider (aside from members of official review panels and the like) who was allowed in. Deep trust was generated only over a few years as a result of (a) never betraying any secrets, (b) working hard to attend pretty well every meeting, thus evidencing commitment, (c) sharing social spaces with the community including eating and drinking (commensality) and such other informal social relationships as opportunity threw up, (d) being completely honest about the nature of the sociological project and developing a website explaining the goals and methods so that there could be no suspicions that he had some surreptitious goals in mind, and (e) working hard to understand the technicalities of the field, including tacit elements (interactional expertise), and displaying this understanding, something which, again, indicates commitment. Point (e) shows that, while in the Checkov case *tacit knowledge was gained because trust was generated*, here, not only is this the case but also *trust was generated because tacit knowledge was gained*.

In sum, from these experiences and the results of other studies of the nature of science we know that:

A) Core sets are small and intimate in organizational style not public, though we know that in big sciences they can be more than a thousand strong.
B) The boundaries of core sets are tightly controlled, either as a result of a self-conscious policy or because interaction with the community depends on the acquisition of technical skills hard-won over the years of the professional apprenticeship.
C) The organization of science is hierarchical but the hierarchy is always aware that the most junior members might have ideas that could promote them instantly to the pinnacle of that part of the structure that evaluates the quality of thinking – the location that those beginning an academic career most want to occupy.[93]

---

[92] Where a core set is too large for everyone to know everyone intimately, one of the mechanisms for the development of a language and the spread of trust will be what we have called 'social diffusion' (See Social diffusion).

[93] This is best analysed under Thomas Kuhn's (1959, 1977) notion of 'the essential tension' in science.

D) Related to dimension 3, in a well-functioning group in science there will be a high level of tolerance for dissent or eccentricity alongside an understanding that science depends on a solidly maintained consensus.

We can call this combination of qualities of local groups and local communication 'core-set values.' The Mertonian value known as 'organised scepticism' (see note 89) is manifested in the core set through the way dimensions 3 and 4 are instantiated, while the value of 'universalism' will show itself in the way the boundaries to entry are potentially open to persons of all types and eccentricities (including that of someone like Collins). What we are arguing, then, is that the values of science are bound up with the ways scientists, as scientists, communicate in core sets when they are making new scientific knowledge. After we have looked at remote communication once more, we will be in a better position to argue that there is something special about small local groups of a certain type that cannot be substituted by remote interaction.

Conclusion

In these first three chapters, we have argued for the importance of trust and socialization in our society. We have illustrated the ways that trust and socialization are dependent on face-to-face communication. Yet, we live in a world where the remote is being injected more and more with the intention of replacing these face-to-face interactions. What potential does remote communication have and where are its limits? That is the subject of Part II while Part III will look at the impact of the change on society as a whole, illustrated by recent events.

PART II

Arguments and Evidence: Can Remote Communication Replace Face to Face?

CHAPTER 4

Remote Technology and Trust

Why remote communication seems to hold promise for the future

We now turn to remote communication. The task of this section is to try to prove a negative: why remote communication cannot do the things that face-to-face communication can do. It is always difficult to prove a negative, especially in a rapidly moving technological domain; perhaps new technological developments will render past criticisms otiose – how can one anticipate the potential of science and technology that has not yet been invented? The same problem faces any critique of artificial intelligence in general: the critic's critic can always say – 'But you said AI would never accomplish 'this and that' but it has accomplished them, so why should we believe you when you say it will not accomplish "the other"?' The critic can only respond by using all the imagination available for seeing into the future – which is why we are happy to look as far as science fiction and argue that 'the other' might be accomplished but it will have to depend on the invention of something currently unforeseeable. We have to demonstrate the negative in as many ways as possible, recognizing that the final conclusion will never be as decisive as we would like, and we have to show why much of the optimism for current technical solutions is misplaced.[94]

We start with the technicalities. Not all remote communication is efficient even where it might be expected to be strong – for instance, even video is poor when it comes to organizing discussion among more than two or three people. Here, however, is an example of recent technological developments that already promise to derail that negative claim! Zoom and similar platforms have found technological solutions to turn-taking that did not exist when the first draft of these sentences was being written – the hand-raising icon and the like. Still,

---

[94] For an analysis of the meaning of impossible in the debate about artificial intelligence, see Collins (2018) Ch. 2, especially the table on p. 22.

How to cite this book chapter:
Collins, H., et al. 2022. *The Face-to-Face Principle: Science, Trust, Democracy and the Internet*. Pp. 81–90. Cardiff: Cardiff University Press. DOI: https://doi.org/10.18573/book7.e. Licence: CC-BY-NC-ND 4.0

we know that these new technological innovations do not deliver all the subtle distinctions and body-language effects that contribute to turn-taking in face-to-face situations – audience interaction with presentations is dull – which is why, for example, stand-up comedians always need a studio audience even if they are recording for television. But assuming the speed and bandwidth of communication technologies continue to increase exponentially, could technology get there? As early as the 1950s Asimov had got as far as 'trimensional viewing', in his imagination at least. But, by now we should be at least alert to the possibility that all the bandwidth we can imagine cannot replace all the advantages of the co-location of bodies, however hard such a thing would be prove decisively.

Then there are more foundational problems. We have already seen that Asimov could not imagine human life in the nursery working with trimensional viewing. Early socialization, and education, from kindergarten onwards, depends on personal interaction because interaction through spoken language seems to be what creates the social world in the first place and the social world makes us what we are. But, as we will also argue, since *we never stop becoming what we are* – continually engaging in acts of specialist socialization throughout our lives – the child's world and adult's world depend on the same deep processes.

But, going back again, technology is full of surprises. New technology is providing ways to generate trust over the internet. We already know that every now and again, people come to trust others through an extended exchange of letters without ever meeting.[95] One of the authors of this book (Collins) has a productive academic friendship with Arthur Reber, who has been mentioned above in connection with their alternative interpretations of what it is to be a cat, but they have never met! The whole thing has been conducted by exchanging some hundreds of emails (we will return to this example). This way of generating trust is unusual, precarious, and takes a long time.[96]

It is true that we are inventing more and more ways of communicating from locations remote from each other and we are also inventing mechanisms which enable us to collaborate in a positive way via websites: consumers write reviews of books and report their experiences of hotels and other services; Uber and Airbnb do background checks on drivers, renters and letters; while the huge, China-based online trader, Alibaba, got off the ground by using escrow accounts so that the seller could be satisfied that payment had been made before sending the goods, while the buyer could be sure that payment would not be discharged until the goods had arrived in the promised state.

---

[95] For example, 'An Epistolary Friendship: The Letters of Elizabeth Stuart Phelps to George Eliot' https://muse.jhu.edu/article/19584/summary. More examples can be found by searching the internet for 'epistolary friendship'.

[96] Baym (2015) Ch. 6 includes a detailed discussion of online personal friendships and how they sometimes mutate to offline friendships and vice-versa; we, of course, are concerned with professional trust relationships.

'Blockchain' technology promises to supply a kind of digital 'trust' between complete strangers. One of the things we have to do if we are to remain alert to the limitations of remote communication is to show that though these things work, they are not a replacement for the face to face, however much their champions claim that they are.

Blockchain technology is the most challenging surprise if one is inclined toward the notion that trust is based in local understanding. Blockchain technology seems like magic – securing trust among any number of people who have never met and almost certainly never will meet in circumstances where violating that trust could result in huge financial gain. Blockchains create records of transactions which are inscribed simultaneously across a network of computers and therefore cannot be changed by any individual; they create a distributed and indelible record of every transaction and every transformation of value so that questions of ownership and who owes what to who can (almost) no longer be doubted. Blockchain technology, as one writer claims, 'will transform how we exchange value and who we can trust' (p. 252) and is one example of the way digital technologies 'can build trust with strangers to connect and collaborate on an unprecedented scale' (p. 258).[97]

Before leaving Blockchain, let us demystify it. Money has a long history, in its early incarnations being a useful medium of exchange because of its scarcity – gold being the principal example in the modern world. But even this kind of material token rests on agreement among humans: for gold to be valuable people have to want gold rather than, say, cowrie shells, and both gold and cowrie shells become worthless when food is running out – as King Croesus discovered. Money, then, is, basically, *agreement* and, given this, it should be less surprising that material tokens like gold can be swapped for intrinsically worthless banknotes, and other promissory notes bearing signatures, and that money can be created by banks at will.[98] Nowadays, in many countries, this freedom to create money is driving inflationary property spirals.[99] Blockchain technology is just a variant of the form of agreement traditionally used

---

[97] Botsman (2017) is a readable description drawn from the management literature of the way digital technology may be transforming trust. She has a wealth of detailed information of this sort driven by her view that society is undergoing large-scale changes in the locus of trust with the internet giving rise to new kinds of trust mechanisms. She sees the guarantor of trust extending over historical time from the local, to the institutional, to the distributed (e.g. p. 262) whereas we are arguing this works only for special subsets of trust.

[98] Ryan-Collins et al. (2012) explain money while pointing out (pp. 139–40) that the State's willingness to accept payment for income taxes in the national currency makes it the dominant player among those whose agreements stabilise a currency.

[99] See, e.g., Ryan-Collins (2018).

to support the value of money, but where the witnesses to the 'signatures' are huge in number because they are distributed across the internet. This makes the agreements especially difficult to question or counterfeit.[100]

Of course, in light of our distinction between transactional and moral trust on (see Chapter 1, Ways of Trusting), it should be readily apparent that Blockchain – and similar services – have a role to play in transactional trust. Does the distributed writing of consumer reviews of books, hotels and services and the introduction of blockchain technology mean, however, that moral trust can now be developed in the absence of face-to-face communication and in large groups as well as small groups? Gaming and hacking aside, the answer is 'yes' in respect of certain limited kinds of *transaction*. These new forms of trusting, because they are based on remote communication, cannot be a matter of moral trust; they have to be transactional trust, blockchain technology (as well as escrow accounts etc.) being a technical solution in the case of remote transactions. So blockchain technology and the rest are not quite the miraculous solution they seem to be since they work only for transactional trust, not moral trust. Moral trust, we argue, is a far more important feature of society, having such a central role in the foundation of knowledge both narrowly and broadly understood.

There is another feature of transactional trust that separates it from moral trust: it can only be sustained if the success of the transaction is visible to all. For transactional trust to work for more than a very short time, the employer will know that the employee is doing the job; the bitcoins will arrive in the account; the driver will have to take you securely to your destination in their car; the landlord will rent you a clean and comfortable property; the tenants will leave the property clean and undamaged; the purchased items will arrive promptly and in good condition; the banks will not impose large and unexpected charges. If trust is not justified, something distinctly different will happen: the employee will slack, the bitcoins will not arrive, the car driver will not get you there or drive very badly; the property will be unsuitable, or the tenant will trash it; the purchased items will not arrive; a trivial accidental overdraft will be followed by a charge based on punitive interest rates. Technical solutions to the remote trust problem, like all transactional trust, depend on everyone being able to see the difference between fair treatment and violations of trust.

This distinction maps onto a similar distinction in the world of science. There is a class of scientific experiment which allows you to tell when they are working because you can repeat them and watch the outcome. If everyone agrees that if

---

[100] Edgar Whitley points out (private communication to Collins, 6 June 2019) that in certain incarnations blockchains can come under the control of 51% of the participants, possibly rendering them untrustworthy if those 51% organise themselves for nefarious ends. https://blogs.lse.ac.uk/businessre view/2019/06/04/blockchain-governance-the-system-is-not-immune-to -capture-by-interest-groups/.

I put together a laser to this design it will produce a beam of radiation that will make concrete smoke, then anyone can confirm it because it is obvious when concrete is smoking. But there is another class of scientific experiments, and a corresponding class of economic transactions of a different kind. In the case of deeply disputed science, no-one knows what the outcome of a well-conducted experiment is supposed to be. For example, in the late 1960s and early 1970s, no-one could be sure whether resonant-bar gravitational wave detectors should see or should not see gravitational waves when they were working properly; either the bars had discovered some unexpectedly powerful sources of gravitational waves or worked in a way that made them much more sensitive than was expected, or the very fact that they seemed to see the waves showed that they were not working or not being analysed properly. In such a case one cannot tell which experiments are doing what they are supposed to do – at least not in the short term – because that is what the dispute is all about. The last vestiges of the argument about what resonant-bar gravitational-wave detectors should be seeing did not fade away until the first undisputed detection of a gravitational wave, using a very different kind of apparatus, was made nearly 50 years later. Something similar applies to business transactions when what they are supposed to do is not completely clear or does not become clear for a long time. An example is investments: though interest might be delivered in the short term, whether they are good investments is a long-term matter. The problem is highlighted in the case of Ponzi schemes where the whole point is that the unsatisfactory denouement is delayed for as long as possible by fulfilling the short-term returns through new investors attracted by the favourable short-term performance.

In the case of both disputed experiments and long-term investments, there are no technological devices that can stand in for moral trust, and supervisory agencies depend on moral trust anyway. In such cases, then, we have to fall back on ways to estimate trustworthiness that are likely, at some foundational point, to involve face-to-face interaction. Estimates of trust based on face-to-face interaction are not foolproof, of course, or there would be no Ponzi schemes, nor confidence tricksters in general, and no failed marriages nor failed friendships.[101] But in such cases, face-to-face interaction seems to be about the best we can do as the basis of trustworthiness.

To avoid unnecessary arguments about how experimental disputes and the like are finally resolved, we have cut through the philosophy and referred to this second way of partitioning the domain of trust (for instance Table 1.2: Trust and reliance in short-term and long-term) simply as the difference between short-term and long-term criteria of success. Note that long-term failures of trust include instances where none of the parties are aware that what they

---

[101] Botsman (2017) describes the misplaced confidence that personal acquaintances gained from their interactions with Bernie Madoff, who drew them into a disastrous Ponzi scheme.

are doing is not to be relied on, as well as examples where they are so aware; Ponzi schemes and scientific fraud exemplify the latter type while good-faith scientific discoveries which turn out to be rejected by the community, or good faith investments that fail, exemplify the former.

As already suggested, trust and communication are like conjoined twins – as explained, communication turns to 'noise' as you cease to trust the person with whom you are communicating. Unsurprisingly, the two kinds of trust are reflected in two kinds of communication – the first conveys 'information', and that works in the case where everyone agrees what should be seen in the short term when trust is justified; the second kind of communication conveys tacit, or cultural, understanding and it is this that can justify moral trust. This is a more difficult kind of trust to attain, and often has very serious consequences and large costs should it fail; we have already used the phrase 'difficult and dangerous truth' and it gives rise to *difficult and dangerous trust*.

What we have done in the analysis of Bitcoin and similar transactional negotiations is show that these do not show that technological advance can replace the face to face, even though they seem to when first encountered. We are never going to be able to prove the negative decisively, but we have shown that one of the more persuasive proofs of the corresponding 'positive' is not what it seems to be. All we can do is press on with more partial demonstrations of this kind.

Tabulating the remote

It is not possible to construct anything as exact as the counterpart of Table 1.1 for the remote because remote communication can be achieved in so many different ways, from smoke signals to anything we and Asimov can imagine, and each method can, in turn, be used in different ways. As far as possible we will list these various different kinds of remote communication in Table 4.1, inserting face to face at the end so as to show where it fits in terms of the table's subdivisions. Henceforward we will be asking whether this or that feature found in Table 1.1 can be reproduced remotely, and vice-versa, asymptotically reaching toward the negative that we will never quite reach.

Table 4.1 lists types of remote communication with various potentials.[102] It is split into three classes, with the main organizing principle being whether

---

[102]   For another way of describing the different kinds of remote media see Baym (2015, p. 6 ff). Baym provides a much more detailed analysis of different kinds of digital media than we attempt here and, *inter alia*, a guide to the wide literature on the matter. Baym provides (p. 58 ff) a rich and scholarly analysis of the difference between face-to-face and remote communication in terms of trust and understanding and how this affects expressed antagonism (p. 64 ff); these are supported by our interview quotations. But we are also interested in the very act of gathering together, which works as a

Table 4.1: Kinds of remote communication.

I	**ONE WAY ONLY**		Locally or widely broadcast talks
II	**POTENTIALLY TWO-WAY**	Rapidity of turn-taking	Publications Letters Smoke signals Fax Email Telegraph Semaphore Social media
III	**POTENTIALLY INTERRUPTABLE**		Telephone/intercom Video-phone etc Social media Virtual reality **FACE TO FACE**

the interaction is one-way or two-way and, if two-way, how rapid is the turn-taking: turn-taking speed increases as we descend the list. We will suggest later that rapidity of response is one of the features that can make remote communication seem more like face-to-face communication and render it more, apparently, trustworthy. The other organizing feature is whether the communication is 'interruptible'; thus, one can interrupt someone mid-turn in a conversation without disastrous consequences whereas one cannot do this with, say, a letter or an email. Interruption might be possible with certain remote media, but the consequences might be disastrous when body language is not helping to persuade the person interrupted that no rudeness is intended.

As mentioned, a complication of Table 4.1 is that different communication methods can be used in different ways. Even humble email can be used in at least two ways. Thus Collins, like many others, uses email as a kind of distant and slowed-down means of conversation and always responds to 'conversational' emails immediately; he likes responses to his emails to come as soon as is practicable; as is intimated above, rapidity and reliability of response engenders a sense of trustworthiness – the other person cares enough to set aside other

---

provider of energy, in which the costs, normally thought of as a disadvantage, are a necessary part, and as a facilitator of other forms of social behaviour such as commensality. As can be seen, our analysis turns on the principles of knowledge-making in society rather than detailed examination of the nuances of conversation. Baym is also interested in the analysis of the current ways of communicating whereas we are interested in the principles of how societies work and potential long-term changes.

tasks so as to pay attention, and is not trying to hide something. Other email users think of the medium more formally and may work through a list of emails when office time is available or even only during working hours, ignoring the inbox during holidays and so forth. Collins finds this extremely frustrating and ruinous to the kind of working relationship he prefers. He finds that these kinds of delays take the energy out of a productive, if distant, academic interchange, making email still less valuable than it could be, compared to face-to-face communication and even less trustworthy in comparison.

Returning to the theme of the advantages of remote communication, we should note that Francis Bacon wrote in 1625 that 'writing maketh an exact man'. At the time of writing this book (August 2019) Collins has just sent, by email, a five-page, strongly critical, review of a book draft written by a long-standing academic friend of his. They followed this up with a two-hour transatlantic Skype conversation in which the obvious trust and friendship that could be exhibited, though based on many previous face-to-face conversations, shared meals, and so forth, (cf. the example of Reber, discussed in Chapter 5, Collins and cross-disciplinary communication), made the exacting exposition of the differences in view manageable. Crucially, the development of the written critique took a couple of days, and none of this would have worked anything like so well if face to face had been the medium for its initial presentation – email (or letter) was crucial. We do not want, then, to return to the stifling and narrow worlds of tribal life: it is just that we do not want to lose the many positive things that come with that kind of interaction either.

Continuing with Table 4.1's list of types of remote communication, 'social media' also covers a number of possibilities – which is why it has had to be included in both classes II and III. In Class II we would find social networks such as Twitter, Facebook, Reddit and Instagram, where users create posts and reply to, or otherwise interact with, the posts of others (e.g. by 'liking', 'favouriting', 'sharing', 'retweeting', 'up-voting', etc). The distinguishing feature of Class III is synchronous communication, including interruptibility, which is a feature of most 'instant messaging' social media – e.g. WhatsApp, Apple iMessage, or Slack – where one user can see that another is 'present' (in the virtual sense) and/or currently composing a reply. Of course, many social media platforms combine multiple forms of communication: Skype, for example, integrates telephone, video-phone and instant messaging; Slack combines the 'post/reply' model of Class II with the real-time messaging of Class III.[103] Furthermore, there is another vital dimension: this is whether the message is permanent or ephemeral. Another of the pressures on face-to-face communication is that once something has been recorded it can be used over and over again, without loss in the case of remote communication, and broadcast to large audiences; this is very cheap and efficient. On the face of it one can also record a face-to-face conversation, but the record is merely a surface image of what took

---

[103] Alun Preece provided the rough classification of social media.

Table 4.2: Some positive features of remote communication.

	FEATURE
1	Immediate or near immediate interaction of remote parties
2	No travel required
3	Relatively environmentally friendly where parties are distant
4	Potential anonymity
5	The exactness of written communication
6	Less chance of control by dominating personalities
7	Broadcast potential (one to many)
8	*Illusion of intimacy*

place in real time so the impression that one can broadcast it without loss is misleading; sound-recording a face-to-face conversation captures no more that than recording a telephone call or intercom communication.[104] One can see how the argument is going to go.

Given these reservations, Table 4.2 lists things that can be achieved by some or all forms of remote communication but that cannot be achieved with face to face. The ones at the start of the list are almost too obvious to be worth mentioning.

The technological complexity of remote communication is immediately obvious because even the first advantage does not apply to letter-writing. The fourth row, anonymity, may enhance the potential of communication where sensitive matters are concerned but sometimes anonymous witnessing or other kinds of discussion might involve travelling to a special location, vitiating the second and third advantages. We are thinking here of the value of anonymity in courts of law or other kinds of sensitive discussions, but the potential for anonymity in internet communications can also be used to disguise the source of opinions or trolling; we will discuss this at length in Chapter 7. The fifth row has been discussed already. The sixth row is the complement of one of the advantages of face-to-face communication – namely its ability to forge consensus from initial disagreement among the parties. This can have the downside where there are dominating personalities; in extreme cases this can lead to suicidal cults and the like. This is less likely where the communication is remote. The seventh row can enhance certain kinds of education and, of course, is the medium of much entertainment, political campaigning and 'news'. The eighth row – the illusion of intimacy – is different: it refers to the way remote communications of the modern, internet or social-media type are especially useful for spreading

---

[104] Collins et al. (2019) demonstrate that transcription of spoken conversation distorts further.

misleading impressions. The illusion of intimacy is the sense that these com-munications involve a trustworthy community even when they could be a set of providers of managed impressions running an organised social media campaign. We will return to these negative qualities of contemporary types of remote communication in Chapters 8 and 9.

CHAPTER 5

Can Remote Replace Face-to-Face Communication?

Remote technology can clearly help with facilitating transactional trust. It can also help to facilitate ongoing relationships that have been based on face-to-face connections. However, as already intimated, it seems unlikely that remote communication can reach the high bar of socialization that is required for forming communities or producing new knowledge. But how unlikely?

Figure 1.2: The fractal model of society illustrates that most of what is learned during secondary socialization is parasitical on primary socialization because the lower ovals in the fractal model depend on what happens in the topmost oval. This relates to the 'Face-to-Face Principle.' It is not just that secondary socialization depends on primary socialization, though it does. The lower domains in the fractal have their own specialist languages – 'practice languages' – so socialization into one of those domains depends on learning the language, and fluency will depend on face-to-face embedding in the mini-society of specialists. That is what is being tested in the Turing Test and the Imitation Game. Furthermore, what happens when people cooperate remotely is that they have already established, not only a common language, but also trust and, perhaps, some specialist elements of reliance in a foundational local setting.

Examples where the success of remote communication depends on an initial period of socialization into a common understanding through the face to face include the international project team responsible for the research which underlies this book, and the group which discovered the first gravitational wave. As explained, the group which wrote this book was gathered almost entirely via email and the group continued to use mostly email throughout the project, with only the occasional meeting. But, crucially, nearly all the participants in the team were linked in a fairly dense face-to-face social network before the project started. As for the first discovery of gravitational waves, the five months of analysis leading the discovery were conducted remotely using emails and

How to cite this book chapter:
Collins, H., et al. 2022. *The Face-to-Face Principle: Science, Trust, Democracy and the Internet.* Pp. 91–104. Cardiff: Cardiff University Press. DOI: https://doi.org/10.18573/book7.f. Licence: CC-BY-NC-ND 4.0

telecons (to which Collins was a party), but this interchange rested on a dense history of conferences and workshops where trust and a common 'practice language' were developed over decades.[105]

It is important to notice that we are talking about *human* interaction. Only humans have a rich language that links their activities across great social distances, while animals, even though domestic animals live and interact with humans, do not engage in the kind of face-to-face interaction that enables them to acquire natural human languages or self-conscious trust. One of the differences between humans and animals is brought out in the debate between Collins and Reber about cats and hallways found above (see Chapter 2, Humans, Sociologists, Psychologists and What It Is to Be a Cat).[106]

Table 1.1 sets out all the things that can be done with face-to-face communication. Table 4.1 classifies remote communication technologies while Table 4.2 sets out some of the unique advantages of remote communication. Now we want to gather such partial evidence as we have in order to explore the question of whether the face to face can ever be fully replaced with the remote. Increasing bandwidth of communication presents a challenge to any claim that face to face will always be necessary. In comparison with email, turn-taking is hugely speeded up by telephone and video, with video probably allowing more interjection and change of direction and more transmission of body language. But though one can imagine a lot of improvement in a two-way interchange and, below, we will give an extended example from our experience, as already intimated, it does not work as well where multiple parties are concerned. The bank of video screens would need to manage subtle shifts of attention-grabbing behaviour that turn on physical movement and facial expressions among all the parties, combining with mutual recognition of the status hierarchy, partly being established at the time, and reinforced by feedback. Many, only locally executable, competences of the human body seem to be needed to signify a shared readiness to give one person rather than another their moment to be the centre of attention. And how would video cope with the value of serendipitous meetings? In the meantime, the Face-to-Face Principle should be borne in mind when we look at where remote communication has been successfully used; remote communication often works as a substitute for the local once initial

---

[105] See Collins (2017).

[106] Responses to this argument include reference to the minimal linguistic abilities of chimps, dolphins, certain birds and so forth, or the claim that 'animals have their own languages'. The existence of some minimal animal languages does not affect the argument about rich languages that embody understandings of human skilful practices; that animals have languages of their own, even if those languages could be said to be rich, would not affect the argument unless those languages could make contact with human groups through socialization in the way that isolated tribes who have 'languages of their own' can make contact given enough interaction.

socialization has given rise to the necessary mutual understanding and bonds of trust and reliance, but simply looking at these successes without considering their local foundations gives a misleading impression – we have to imagine the long-term consequences of these shifts, long enough for pre-existing trust relationships to have no input.

One example of this kind of misleading impression is found in a study which is famous among the management community since it purports to describe a mechanism by which human abilities can be transferred to machines – in this case a commercially successful bread-making machine. Nonaka and Takeuchi claim that the bread-making machine transfers the skill of the craft-baker into machinery but more careful analysis shows that the success of the automated bread-maker depends (a) on the existence of a network of readily available materials and skills that have been long taught in the kitchen even if not directed specifically at making bread, and (b) on another set of 'instruction-understanding' skills that are also taught through long apprenticeship; this success at remote execution of a skill depends, once more, on the Face-to-Face Principle.

The same goes for the TEA-laser case described above: since the early days, the whole physics community has learned to believe that the laser will work when properly put together – a belief that is, as we have seen in the Checkov case, essential if a difficult skill is to be brought off. The physics community has acquired certain parts of the tacit understanding of how these things work through their general education – perhaps the importance of inductance in very high voltage system. The physics community has been put in a position to acquire turn-key components, or even the whole laser, which has been designed to work reliably, not by incorporating skills but by converting elements dependent on tacit knowledge into elements that are explicitly understood, particularly where relational tacit knowledge is concerned.[107] To think the immensely complex process is simply a matter of 'translating the tacit into the explicit' is the kind of pitfall to be avoided if the relationship of the face to face and the remote is to be properly understood. We now complete this chapter by describing some experiences that bear on these questions; having them in mind for some time, Collins has been reflexively examining a substantial proportion of his communication practices for some years – a kind of personal reflexive ethnography – but the account has application across our society, so provides an invitation to others to re-examine their own interactions and confirm or deny what is being suggested here.

---

[107] For the TEA-laser case see Collins (1974) and Collins and Harrison (1975). For bread-making machines see Nonaka and Takeuchi (1995) and Ribeiro and Collins (2007). For the replacement of 'polimorphic' by 'mimeomorphic' actions and how this relates to tacit knowledge see Collins and Kusch (1998) and Collins (2010).

Collins and trust among the gravitational wave community

We start by examining Collins's earning of trust among the gravitational wave community. Earlier in the book (See Chapter 3, Small face-to-face groups and science) the way Collins gained enough trust to be given access to the very secretive gravitational wave core set was discussed. The list is repeated here:

(a) never betray any secrets
(b) work hard to attend pretty well every meeting, thus evidencing commitment
(c) share social spaces with the community including eating and drinking (commensality) and such other informal social relationships as opportunity throws up
(d) be completely honest about the nature of the sociological project – developing a website explaining the goals and methods so that there could be no suspicions that there was some surreptitious goal
(e) work hard to understand the technicalities of the field, including tacit elements (interactional expertise), and display this understanding, something which, again, indicates commitment.

Would these trust-inducing actions be manageable with remote communication alone? We set the suggested answers out in Table 5.1 for easy reference.

The social scientist can achieve a, b, c and e only via F2F interaction. So this is a small piece of experience suggesting that it would be hard or impossible to replace the trust-gaining elements described with remote methods; it squares

Table 5.1: What elements of face-to-face communication can and cannot be replaced by remote communication?

CAN REMOTE COMMUNICATION REPLACE FACE TO FACE COMMUNICATION?	
a	Not betraying secrets could be managed remotely but, as an outsider, one would not learn any secrets either to betray or not to betray if communication was entirely remote
b	Attending meetings so as to demonstrate commitment would not be manageable by definition as that requires actual attendance and the time and effort commensurate with it
c	Commensality and other informal social gatherings are also F2F by nature
d	Sustained honesty and a website *could* be managed remotely
e	Gaining technical understanding of the tacit seems to be attainable only F2F though this does not take account of science-fiction type virtual reality but only so long as trust could also be conveyed in some way

with the responses of the scientists Collins interviewed at the Budapest conference in 2015 and which have been quoted earlier in the book.[108]

Collins and cross-disciplinary communication

Again, because the problem of remote communication was always in mind, we can re-examine a prolonged but fortuitous remote interaction between Collins and well-known social psychologist, Arthur Reber, who was mentioned above (see Chapter 2, Humans, sociologists, psychologists and what it is to be a cat). It enables us to compare what can be done, and some of the things that cannot be done, with purely remote communication and to make a comparison between email and Skype. Reber is a well-known American professor of psychology who wrote a paper with Collins published in 2013. But Reber and Collins had never met nor read each other's work, and the paper was based on around 600 email interchanges over three months.[109] Reber and Collins have continued an episodic email correspondence, counting themselves as 'email friends', and by autumn 2018 must have exchanged about 1,000 emails altogether. Psychology and sociology have an academic relationship like that of cats and dogs, and that first paper was an exploration of why they had never read or cited each other's work, given that a central theme for both of them was 'implicit learning', in Reber's terminology, and 'tacit knowledge' in Collins's. These terms connote a similar interest, though as the initial email exchange unfolded it became clear that they actually viewed the world of the phenomenon in different ways and to that extent that there was not much to be gained from drawing on each other's work. In other words, in this interchange they were attempting to successfully navigate a trading zone between disciplines, something that ought to have been possible only through face-to-face socialization. Because of the attention paid to this problem over a long period, its salience in the interchange was enhanced and the way it was manifested can be reported.

---

[108] The results are reported in full in Collins (2018), Ch. 8.

[109] Collins and Reber (2013) came about because Reber acted as a referee for one of Collins's papers. Collins asked the editor of the journal to ask the referee if they would be willing to talk directly over a particular point of disagreement. Reber agreed and also agreed to Collins's subsequent suggestion that they write a joint paper concerning the different meanings and approach to the topic of 'tacit knowledge' found in psychology (Reber) and sociology (Collins). Collins had noted that they had both worked on the topic all their professional lives yet neither had ever cited the other. Reber recently invited Collins to contribute a paper to an edited collection which he was putting together entitled *The Cognitive Unconscious*.

The email interchange

Two things are notable about the initial exchange; first, it was occasionally very frustrating as each party failed to understand the other, and it came close to collapsing once or twice; secondly, and more surprisingly, both parties did manage to learn something of each other's world view and even incorporate elements of it into their own – Collins can attest that this was very important for the most recent twist in his own world view – something that is explained in more detail in Chapter 7 of his 2018 book on artificial intelligence. Nevertheless, Reber and Collins remain divided by disciplinary approach and cognitive interests. They are not committed to each other in the way that, say, the members of an experimental team are committed to each other, with each person's time being well spent only if the other team members are spending their time with integrity. The main motif of the interchange was good will and a distant friendliness hampered by frustration. The problems were summed up on pages 150–53 of the joint paper itself, in a series of seven points meant to act as a warning and guide for others attempting to talk across the psychology–sociology divide and other pairs of distinct disciplinary domains. The question is the extent to which these problems could have been resolved by, or alleviated with, face-to-face conversation, and we now add some speculation covering this question.[110] We will ask whether it was the remote communication that was the problem and, in italics, whether face-to-face communication (F2F) could have resolved all or any of these seven problems.

Seven problems of communication: an interdisciplinary email conversation where the parties did not already know each other

The first difference is related to individual thought style rather than clashing disciplines:

1) **Mismatched thought styles**: Reber generally sees continuities whereas Collins is drawn to bringing out sharp differences and classifications. Reber focuses on individuals, Collins on collectivities. *It seems likely that the sheer efficiency of F2F would have created time for this kind of difference to be discussed up front as part of a reflective 'meta-discourse' on the main substance of the discussion; this may have resolved some of the tension.*

---

[110] An important analysis of the problem of cross-disciplinary communication starts with Kuhn's notion of paradigm incommensurability. Our duck-rabbit story is a way of explaining it. An analysis of the various ways of communicating across disciplinary boundaries is found in Collins, Evans and Gorman (2007) and extended in Collins, Evans and Gorman (2019).

The second difference is typical of strain between disciplines:

2) **Semantic mismatch:** The parties often use the same word but with different meanings without realizing it. The problem applies even to words at
the very centre of the discussion: as we mentioned earlier, for Reber 'tacit'
is a synonym of 'unconscious', for Collins, 'unspoken' or 'unsaid'. Reber
feels he is being loyal to Polanyi [the instigator of the term 'tacit knowledge']. Collins notes in TEK [his book, *Tacit and Explicit Knowledge*] that
he is deliberately going beyond what Polanyi intended. In some instances,
we were agreeing on the nature of tacit knowledge; in others disagreeing – and were often bewildered by the incoherencies that emerged. This
kind of disconnect is dangerous because the terms are so familiar that it is
hard to imagine they might mean something different to the other party.
*Again, there is much more space and opportunity for meta-discourse in F2F
and this might have helped.*[111]

The third difference is both individual and disciplinary:

3) **Mismatched explanatory adequacy:** As Collins sees it, the parties justify
claims in different ways. We saw this in a discussion of whether machines
can be classified as potentially knowledgeable or non-knowledgeable.
According to Collins, Reber works by building a consistent world view
based on consciousness as inherently subjective and embedded in evolutionary theory. Collins claims there must be observable consequences [of
a theory] and Reber does not have them. Reber thinks Collins is wrong
and that he does have evidence for his position that is as strong or stronger
than the evidence for Collins's position. But the notion of mismatched
explanatory adequacy remains useful even if, as in this case, both parties are wrong about what the other is trying to do. There are many cases
where parties disagree about what kind of grounding is needed to establish a scientific result. *In this case, F2F probably would not have helped as
we explored the problem pretty thoroughly in the email exchange without
being able to resolve it. Collins has encountered the same mismatch of expectations in discussions with others who see themselves as loyal to Polanyi.
Collins's position emerges out of thinking of the question of intelligence as
arising out of the Turing Test or other comparison tests. Those influenced by
Polanyi, or questions of consciousness, argue about the internal states and
processes upon which human intelligence is based.*

---

[111] That said, the example of Collins and Kusch struggling over the meaning of
the term 'action' (see Chapter 2, Humans, Sociologists, Psychologists and
What It Is to Be a Cat) shows that F2F is far from an automatic solution to
this kind of problem.

The fourth difference is typical of where the parties are not jointly socialized:

4) **Mismatched saliences:** Negative mismatched saliences occur because to remedy an information deficit, one needs an inventory of what is in the other party's head so one can see what is missing and try to fill the gaps: such inventories do not exist. Positive mismatched saliences affected the exchange because one party continually explained at length what the other party already knew. For example, Reber explained the psychological equivalent of Dreyfus's five-stage theory of expertise several times because Collins's ignorance of it as he saw it [Collins knows it thoroughly, knowing Dreyfus personally for many years] seemed, to Reber, the only way to make sense of aspects of Collins's position. Positive mismatched saliences are very frustrating as they stop debates moving forward but they would probably be less marked in face-to-face conversation. *As far as positive mismatched saliences are concerned, F2F would probably have been of great help here: Collins can imagine himself expostulating 'don't tell me that all over again, I've already told you I know it inside out', in a way that would have been impactful but far too rude for email. This kind of remonstration might have led the exchange in more productive exploratory directions. With negative mismatched saliences it is not so clear that F2F would have helped since neither party would have known what to say to help things along. But it is possible that things might have become clearer with F2F just from the way words and silences were distributed in the whole corpus.*

The fifth difference is typical of disciplinary difference:

5) **Focus blindness:** It is sometimes impossible to see a contribution that lies in the peripheral field of a strongly focussed gaze. On occasions one of us thought he had asked a certain question, but the other did not see the question because it was outside his view of the scope of their project. The exchange would continue on the assumption that the other had seen and appreciated the contribution. Confusion followed. *Once more, it is hard to say if F2F would help here, but it might.*

Reversion is a general problem too:

6) **Reversion:** Often one of us would explain an effect X to the other whose response made it evident that he understood. But the understanding was temporary. The problem was that the understanding of X was held together in the longer term by a supporting semantic net in the discourse which included W, Y, Z, etc., and the whole structure maintained its integrity only through continual use. These Ws, Xs, Ys and Zs are like the spinning plates in a juggler's act – if they are not kept spinning, they fall. The dialogue, which is continued by one of the parties as though X is still

in play, reverts to the earlier state of mutual incomprehension. *The sheer density of conversation in F2F and the readiness with which points once made can be repeated – something much harder in a written conversation where repetition seems like wasteful redundancy – would surely have helped to keep the plates spinning. If we had brains like computers, with indefinitely large, permanently stored memory, we might be able to hear something once, take it in and keep the plate balanced – and therefore keep a million plates balanced – but, as it is, we humans can keep only a certain small number of plates balanced and only with continual spinning-up. That spinning-up is the use of words in conversation of a community. That is one powerful way in which socialization works. Even once learned, any prolonged distancing from the community leads to the language deteriorating – the tacit knowledge becomes degraded. Once more, embedding in groups is the key.*

The final point is a problem associated with any academic argument:

7) **Misplaced engagement:** Often, to explore and explain two cross-cutting views of the world, one needs to be distanced from them. But because we were engaged in the worlds we were trying to explore, it was almost impossible for us not to slip, every now and again, into trying to convince the other that they were wrong—the traces are still there in the published paper. Where possible the argument should come only after the mutual exploration. *F2F might have made this kind of problem more transparent though that might not have resolved it.*

As the longer email conversation went on over the years, it became apparent that the parties would not have been able to work together because they disagreed about too much that was fundamental, notably Reber's exploration of consciousness and his writing a book about consciousness, none of which Collins found convincing or persuasive. Collins watched a video of Reber giving a talk and answering questions at a conference and had to tell him, by email, that he felt his answers were inadequate and seemed to him not to face up to the challenges being posed. Collins felt that in this he was pushing the envelope of what could be said via email. Email is harsh because, as we have discussed around Table 1.1, criticism comes across too sharply, not being moderated by body language – the smiles and the like which can tell the person you are talking to that you admire them even though you do not like some particular aspect of their work.

A comparison with Skype

In September/October 2018 Collins re-opened contact with Reber because he wanted to learn more from this experienced social psychologist about the topic of this project – face-to-face versus remote communication. One of Reber's friends, who Reber solicited, did indeed send a series of useful URLs dealing with the organization studies literature. On another topic, Reber sent to Collins a jointly authored paper of his outlining a critique of parapsychology, remarking that he thought Collins would not like it. Collins was very disappointed with the paper, one of a genre with which he was familiar, and felt that he should tell Reber so in no uncertain terms, but felt that following up his criticism of the consciousness material, this would be stretching an email friendship too far. Therefore, Collins decided to experiment with offering this risky criticism to Reber using the medium of Skype instead. Inter alia, this would test whether the video medium would transmit enough moderating body language to make it possible to be as critical of this paper as he wanted to be in spite of the preceding negative context. Thus, after about five years and about 1,000 emails, Reber and Collins were to 'meet' for the first time – via a video link.

At that time, neither party were practised video-link users or enthusiasts, Reber having long disconnected his Skype facility and Collins using it only for weekly meetings with grandchildren and almost never for academic interaction (Collins now uses Zoom frequently and positively). Nevertheless, the 'meeting' was surprisingly successful, with Reber and Collins being obviously delighted to 'meet' each other for the first time after so many years. The delight was even more clear given the difficult circumstances: the time-zone disparity meant the Skype had to happen at 10.30pm for Reber and 6.30am for Collins. Fortunately, there were no delays or other technical difficulties and both sound and pictures were good, without pixilation. Collins recorded the 78-minute interchange on a sound-recorder, an attempt at video-recording not being successful.

The personal atmosphere is indicated by the fact that Collins would have asked his wife Susan to come to say hello to Reber if she had not been away at the time: this is something you cannot do with telephone – you cannot 'meet someone' over the telephone though you can over video – just about. Note also the importance of a sense of a first meeting between Reber and Collins and a lot of affirmatory smiles – we were pleased to see each other even though harsh things were going to be said. (Note also that Collins discovered he had been mispronouncing 'Reber' all these years as 'Rebber' whereas Reber's preferred pronunciation is 'Reeber'; thus is text different to sound.

Later Collins did some simple analysis of the first 40 minutes – approximately half – of the recorded exchange – these are consistent with the turn-taking we described that happened in the first face-to-face meeting of the project that is the basis of this book (the Reber exchange actually took place first).

In this first 40 minutes of the Reber-Collins meeting, there were 79 double turns which included 63 affirmative or negative interjections which did not halt the flow of a turn. This is almost certainly not as efficient as a full face-to-face conversation but is still stunningly different to an email exchange. If we assume there would have been about 160 double turns in the entire 78-minutes that is equivalent to about a month's worth of our paper-writing interchange. Of course, some of those email turns would have been more elaborate and more carefully worked out – more 'analytic' – and might have contained references and the like so we are not comparing like with like, but in terms of reaching a *mutual understanding*, that kind of analyticity is not necessarily helpful; continual interjections, corrections and iterations seem more promising. Also, non-halting interjections are almost impossible over email, where one turn has to be completed before another is embarked upon. The interjections in more normal conversation are part of the 'meta-conversation', indicating when one party thinks they have understood the other or agree with them, or when they show they have not understood or agreed – they show almost instantly whether the other party needs to adjust the flow so as to expand on a point or stop expanding on it because it has been said to be understood which moves the point on; they also help to keep the necessary plates spinning (point 6, 'Reversion', above).

Utilizing Skype instead of email, Collins did, however, feel able to put his objection to Reber's parapsychology paper in the strongest possible way, expressing it in terms of Reber wasting the remaining years of his life on such a misplaced project. In retrospect, putting it this way expressed disgust with Reber's academic choice while simultaneously expressing a concern with his well-being, but it could only be done in a semi-humorous way, while it would be a huge risk to imagine that this element of humour would come across in email. Incidentally, we continued to exchange emails about a particular point of substance in the article – or, more accurately, Collins continued to badger Reber about it – but the sense of good humour may not have been robust enough and Reber soon made it very clear that this line was futile. The underlying disciplinary and philosophical/meta-methodological divergences are still preventing agreement at the time of writing – whether such deep disagreements could be resolvable between two individuals, even in face-to-face communication, is far from clear.[112]

So, we can see from the Reber experiment that Skype is more efficient in terms of turn-taking than email, but face to face is more efficient still. But the experiment does not tell us anything decisive about whether face to face will more readily lead to agreement because we all have experiences of face to face leading to disagreement. On the other hand, once we have gained all we can from face-to-face discussion we know, for sure, where we stand, whereas even video can leave us wondering whether more intimate discussion would have been more productive in terms of the advance of scientific understanding.

---

[112] It is a pity the parties live 6,000 miles apart because a proper face-to-face conversation would make a still more interesting experiment.

The future of corpus analysis: word embedding set free

We now move away from experience and speculate, once more, about the future. As intimated, new technologies and increasing bandwidth are always going to be a threat to what is argued here about the importance of the face to face. We cannot prophesy but we can try to extrapolate.

In an earlier part of the book (see Chapter 2, Duck-rabbits and the bath of words) we looked at the way meaning might come from the relationship between words and silences in entire corpuses of speech and that, though it was early days, computer analysis of this 'word embedding' shows some signs of uncovering this meaning. With the power and storage capacity of computers increasing exponentially, not to mention the rapid increase in the use of surveillance technology, we can imagine every word heard by Collins in the course of some future counterpart of his 45-year-long fieldwork on gravitational waves being stored and available for analysis.[113] We might be able to trace the changing reality of detected gravitational waves in the course of those 45 years in the changing relationship of words and silences in the entire corpus of gravitational wave-talk in interviews, and conferences in corridors, coffee bars, and over lunches, dinners and drinks. Collins's analysis is a record of the changing cultural meaning of gravitational waves over that period, but the purely statistical and meaning-free analysis of the corpus of words in which he was embedded might turn out to give rise to the same understandings as he developed complete with all the interactional expertise he acquired, which includes the tacit knowledge associated with understanding the practices – not being able to carry out the practices (that needs a body), but understanding the practices to the level, of say, a technical manager. Maybe a colossal statistical analysis of the relationship between words and silences in the entire corpus of gravitational wave talk to which Collins was exposed would be the equivalent of the work Collins did as an immersive social scientist. And we do not have to stick with Collins, we can imagine the gathering up and storing and analysing of what everyone hears and says about gravitational waves.

If this statistical approach was successful, we could compare an analysis of the face to face with an analysis of the language found in the published literature and, of course, email and videos and so on. We could statistically investigate whether what you get from F2F by being bathed in language is the same as what you can get from remote. If we accept the idea of interactional expertise, then culture, including its practical aspects, must be captured in language and maybe it could be abstracted from the two different kinds of communication and compared; to find out we would need a big enough database of conversations, and other words, because all these results are parasitical on the corpuses that are extracted from society for analysis. Maybe the equivalent will one day be managed for national cultures, not just developing sciences. Whether

---

[113] See Collins (2019) for the method.

a society that was ready to capture, record and store all that spoken language and other communications would count as a utopia or a dystopia is not so clear – it would need lip-reading computers on every street corner not to mention the home – but the idea of an exploration of how language and culture relate remains fascinating.

Note, once more, that the way language develops depends on the practices of a society – to repeat an earlier example, without tennis players there will be no talk about tennis to analyse – but the way the language captures practices, once the culture and its practice languages are formed, is another matter. We can imagine this data being gathered and we can imagine these analyses being done and we can imagine that they might prove or disprove what is being argued here – that embedding in face-to-face conversation is vital if the implicit cultural meanings in the distribution of words and silences is to be acquired by social actors.

Even if we had massive and representative corpuses of language stored, regularly updated and available for analysis by deep learning techniques, would this truly enable us to explain and/or reproduce the way humans learn from the bath of language? Today's experience with deep learning computers tells us that it would be incomparably better than anything that has been done before the huge data bases and deep learning analytic technique came along. But there would still be no body language – no ability for the machines or the analyses to see the difference in meaning between a provocative remark expressed with a smile or other warning sign and one not so accompanied because these things do not appear on the transcript of speech (that is why we use emoticons and emojis in a weak attempt to add some body language but it is *we* who add them, not the machine or the email interface because it is only we who understand what they are meant to represent).[114] So there is still reason to think that while immersion in the bath of language conveys a huge amount of what we learn through the process of socialization, it is always going to be incomplete unless the touches, nods, winks, travel, consumption of food and drink and the commitments that go with such consumption that accompany face-to-face conversation are captured too, and that is something we do not know how to do. That is an attempt to get as near as we can get to the proof of a negative based on as much imagination as we can muster.

The crucial difference between information exchange and socialization is the ability to use language in an innovative way. What that involves is knowing the difference between the two kinds of linguistic rule-breaking: the kind that is a matter of making a mistake and the kind that makes language interesting. Here it is, illustrated again: (a) 'my computer always flags the misspelled word even when it should not be flagged, as in "I'm going to misspell wierd as an example"'; (b) 'my computer always flags the misspelled word even when it

---

[114] Efforts to achieve machine understanding of emojis are, nevertheless, an active area of research, e.g. https://deepmoji.mit.edu.

shouldn't be flagged, as in "I'm going to misspell wried as an example'". There are two misspellings of weird, one of which is interesting and should not be flagged and one of which is just a mistake; it is a mistake about how to misspell properly in this context. I know this through my linguistic socialization. Can I learn that difference from analysing corpuses? It is hard to see how.[115]

This excursion into the future is going to fail in the face of increases in bandwidth which we might call 'super virtual reality' – a kind of virtual reality that transmits smells, touches and other local sensations as well as the holographic viewing imagined by Asimov. Could this kind of virtual reality transmit the sensation of being in a small local intimate group to many thousands at the same time and, the communications having been recorded, could it transmit it again and again to new groups? It is hard to say without really understanding all the things that go on in genuine F2F such as are set out in Table 1.1. Could it transmit the equivalent of the touches and warmth of a parent along with the responsiveness to sounds being offered to a new-born and all the multiple experiences that crowd in as the child gets to nursery age and becomes an adult including serendipity, the commitments expressed through commensality, and the energy that comes out of the shared costs of arranging face-to-face meetings, the very costs that virtual reality is designed to circumvent? This takes us too far into fantasy-like rather than anthropology-like science fiction, but, in any case, near-infinite virtual reality will not be here for a very long time, so the book's analysis, if it is right, will still be valuable for the world as we will know it for many decades, even if there is a very long-term science fantasy future to come.

---

[115] This whole argument is developed at length in Collins (2018).

CHAPTER 6

Small Groups to Big Groups: When Big Groups *Are* Trustworthy

We have argued in Part I that face-to-face communication is the key to developing trust, reliance and socialization in general, and we are arguing, in Part II, that remote communication cannot, for the foreseeable future, replace face-to-face communication in small groups in these respects. We have explained that there are ways of developing trust via remote communication, but, odd counter-instances aside, such as Collins's relationship with Reber which did not create or change cultures in any deep way, we have set these to one side as working with *transactional* trust rather than *moral* trust; we have argued that the new developments will not work for 'difficult and dangerous' moral trust. But the conclusion will still be precarious if we cannot answer an old question: How it is that modern societies work at all? In modern societies, so much involves trust in strangers in remote locations, that the old question is even more puzzling in the light of our analysis. As we have argued from the outset, the basis of society is trust and reliance born of face-to-face communication in small groups. That is the topic of this chapter – we need to find out if the very existence of modern societies renders our whole approach otiose. In Part II we are asking: 'Can the remote replace the face to face?' If it has already done so, we already have the answer, and it is not the 'no' that we have articulated so far. This short chapter is intended to resolve this potential obstacle to the Face-to-Face Principle.

In the modern world, as we grow out of childhood our families teach us to go a little beyond the distrust of strangers and to trust certain larger institutions, perhaps the police, or perhaps not, perhaps a religious group or perhaps not. Then we learn to trust or rely on certain systems of transport, restaurants, the government, banks, the health service and so on. We will argue later in the book that we need to move societies' trust instincts more in the favour of the institutions of science and similar groups of experts, but the question we are

How to cite this book chapter:
Collins, H., et al. 2022. *The Face-to-Face Principle: Science, Trust, Democracy and the Internet*. Pp. 105–114. Cardiff: Cardiff University Press. DOI: https://doi.org /10.18573/book7.g. Licence: CC-BY-NC-ND 4.0

asking here is how we ever could have learned to trust such wider institutions in the first place? Forgetting the modern technological fixes that, in any case, work only for certain classes of transaction, how did trust ever spread beyond the local community, allowing large societies and nations to form, and how do we maintain trust in anyone who is not part of our local group?

As intimated in the phrase 'illusion of intimacy', we will argue that something of major significance can be revealed by looking in reverse at the idea that trust begins in the family and the local group. We will argue that dangerous things happen when we confuse people who are not part of our local group for locals. George Orwell's *1984* captures the point with the dictator 'Big Brother' whose very name invokes trust in a benevolent family member. But still more relevant today, as we argue in Part II, are certain forms of social media, seemingly intimate forms of communication between personalised interfaces with identities, but which are actually mass communication, adopting the form of the personal and giving rise to the illusion of intimacy. We live our lives trying to work out who to choose among those competing for our trust, and the traditional and relatively reliable choice has always been the local, or those who themselves trust the local, but with the way social media is being deployed, the distinction is becoming more and more confusing;[116] the new technological developments are another confounding feature that we look at in the next part of the book.

Non-personal, remote trust

Turning away from communication for the moment, it is not simply the local that is justifiably trusted or we would never go on holiday, never take a taxi or get on a bus in a populous town, especially abroad, nor would we leave our money in the bank, buy anything the value of which was not immediately evident, especially if it was mechanically complex, nor of course, would we purchase anything online or trust a professional to provide a service for us. It is easy to imagine what a world without routine trust and reliance would be like, since even the world we live in often turns out to be untrustworthy, where certain goods and services are concerned. Buying a horse, or today's equivalent, a used car, has always been an icon for a transaction fraught with peril. Sometimes we try to displace the problem to a professional agency who will check the car for us. Likewise, we try to displace the problem of assessing the soundness of a house to a surveyor.[117] Nowadays, in Western societies, as mentioned, there are new dangers associated with putting your money in a bank both from

---

[116] This transition, from taken-for-granted reliance to self-conscious trust, is what that underpins Anthony Giddens' version of reflexive modernity (e.g. Giddens 1990).

[117] But why do they wear smart suits and never crawl into the dirty roof space (at least in the UK)?

unexpected interest charges on overdrafts and of losing everything should the banks' gambling bankrupt them once more as it did in 2008 (which is why, in the UK, the government has to guarantee deposits up to a certain level). In the film and book, *The Big Short*, we are told that the securities-guarantee firms, who continued to give subprime mortgage investments the highest 'AAA' ratings long after defaults had gone through the roof, were doing it so that the banks would continue to buy their services and, in turn, big-bank owners of the worthless stocks had time to sell some of them – a collapse of that long-range trust the continuity of which seems, as a result, even more of a puzzle.[118] An economic ideology which equates self-interest with efficiency, and obscene financial rewards with evidence of social worth, seems to push what were once honourable professions more and more in this kind of direction, with politicians and economists alike seemingly unable to understand the extraordinary economic efficiency that goes with people being willing to work for the intrinsic rewards of their profession alongside just payment, rather than for financial incentives alone. These politicians and ideologues hasten to sacrifice the delicate fabric of trust and the sense of vocation, which are the basis of societies, and replace it with the far less efficient management control of public services, and financial incentives in the private sector. Collins and Evans argue that only professions such as science, where integrity is an intrinsic and defining characteristic, have, at least, the potential to isolate themselves from these pressures.[119]

Nevertheless, Western society has not yet collapsed, and people still do take holidays and make mid-level transactions as a matter of course. Somehow, given that we have set out the view that trust begins in local groups, can we explain how it works across great distances in modern societies of strangers and even as far as foreign societies? We will need to understand this if we are properly to understand trust in small groups.

When the remote is really local

Straightaway, it is important to note that the problem is not always quite as difficult as it seems since, actually, much of our trust in distant strangers is really a form of local trust working indirectly. When we book our summer holidays in remote locations and book our airplane flights and all the rest, paying for all this in advance, perhaps over the internet, we are often trusting our local acquaintances who have done it before and, through their stories and silences, and willingness to repeat the experience, are telling us that all is safe and secure and that our payments will be honoured: effectively, we have sent local people we trust to explore distant lands and distant transactions for us, and they have come back with positive reports on which we base our actions. Think

---

[118] *The Big Short* is Lewis (2010).
[119] *Why Democracies Need Science* – Collins and Evans (2017b).

about it: not so long ago Vietnam was one of the most dangerous places on earth for a Westerner, but somehow it has metamorphosed into an attractive holiday destination: for those of us who were not pioneers, it is the reports of travellers who we know that allow us to give it serious consideration; these reports are far more reliable than those of the securities-guarantee firms *because* they are offered freely, not purchased. The occasions when such trust is not honoured make news, and that shows just how rare they are – as rare as 'man bites dog' – and also why it is news when there is a breakdown in, say, travel arrangements and the question arises of whether agencies that have been established to, effectively, insure travellers against losses, are brought into play.[120]

This may also resolve the problem of how we know that we cannot reserve an empty seat on a bus, an example set out to exemplify 'sociological metaknowledge'. The story concerns a conversation that took place in the Livingston site of the Laser Interferometer Gravitational-Wave Observatory (LIGO) – about 40 miles from Baton Rouge, Louisiana – around the year 2000, between Collins and Gary Sanders, the project manager of LIGO. They were comparing sociology and physics in terms of method.[121] Collins justified sociological methodology by claiming that both he and Sanders were more certain that you could not board a city bus in Baton Rouge and ask for two tickets, one for you and one to reserve the seat next to you and keep it empty, than they would be about the first detection of gravitational waves. (The claim would be confirmed about 15 years later.[122]) The question is, where does that certainty about buses come from, given that neither Sanders nor Collins had ever used public transport in Baton Rouge? The answer could be that sociological metaknowledge is a form of indirect local knowledge generated in a similar way to knowledge that comes from sounds and silences in conversational settings. Just as in Collins's conference-day in Pisa (see Chapter 2, Silence and sounds in Pisa), when he

---

[120] And in the instance of the Covid pandemic, fail to honour their promises. The example of international travel by aeroplane also appears in Giddens' (1990) discussion of modernity. In that work, Giddens argues that trust in abstract systems has become an essential and unavoidable part of modern life and that the 'facework' of human employees – the representatives of the abstract system – plays a role in justifying that trust as they provide the means through which customers interact with the system. We agree that this is part of the story but, as argued in the main text, we believe it omits the more significant contributions made by local, face-to-face networks.

[121] A more complete account can be found in Collins (2019). A solution to how we have sociological metaknowledge is not offered in that source – we are trying to solve it here.

[122] Since, as only the insiders knew, there was some nervousness about the 14 September 2015 first detection until a second event was seen, on 26 December 2015; this second event was not made public until June of the following year, long after the confident announcement of the first event.

hardly heard Joe Weber's name mentioned and that gave him knowledge of the status of Joe Weber, Collins has never heard anyone say that Baton Rouge, or any other part of the US, is a bit different from other places when it comes to buying bus tickets. So, from the fact that he has not heard it said, it is a reasonably secure assumption that the organization of bus companies is pretty similar to what it is in his own locale – and likewise for Sanders. Our sociological metaknowledge, our knowledge of how far our knowledge of how things work extends outward from the local in the case of any particular social institution, is really just a part of our *local* cultural understanding. Our local understanding embodies an implicit understanding of what happens elsewhere brought home through the experience of inadvertent explorers and ambassadors diffusing back into our local world even though they may never actually say anything; the silence in our own locality informs us that things 'over there' will be found to be familiar because if they were not we would have heard about it; somebody *would* have said something. It is not so clear that the same applies to remote interaction because one has to ask questions actively to discover unusual pieces of information like that – why should people be talking about buses in Baton Rouge over the internet, though it would be perfectly natural for people to chat about after a return from a holiday in such a location?

The reason that every excursion from one's front door is not a fearsome adventure beset by doubts about what kind of people might be encountered in the next street, and the reason one does not fear to get into a taxi cab driven by a complete stranger, is that if these things were dangerous one would have heard about it just as, when a child, one heard from one's mother and father that strangers *were* dangerous, as, indeed, they are to small children. So, we can resolve some of the problem of trust in the extended world by noting that it is really something we learn about from people who are trusted because they are local.

In the original exposition of the Baton Rouge story, it was said that this sociological metaknowledge about bus-riding in distant locales did not extend as far as Pyongyang; now we can see why! Pyongyang (at least in the 2010s) has very few run-of-the-mill visitors from the West, just enough to have inserted into our local knowledge that it is a strange place but certainly not enough to have provided any input to the local culture about how buses work, either through positive statements or silences. For this kind of silence to mean something, we have to know that the potential is there for the silences to be replaced by sounds – accounts of something 'funny' going on – and in the case of Pyongyang that potential is not there. There are then, three kinds of silence: silences that mean something because they upset the normal rhythm of conversation, and this other kind of silence that can be separated into two classes: the meaningful silences that indicate there is no problem and the meaningless silences which arise out of lack of knowledge. Another part of implicit (tacit) sociological metaknowledge is knowing what kind of thing it is that we are not hearing.

Social diffusion

Once more, language is at the heart of the kind of sociological metaknowl-edge that has just been described. But in modern societies language diffuses across the whole society and links many domains. Language bridges the local and the remote! We will call the mechanism that enables this 'social diffusion'. Social diffusion uses personal interaction as the medium, but the result dif-fuses beyond the co-located group. Natural languages continually evolve in response to inputs that are distributed across whole societies, and the changes spread across whole societies so that the language spoken by widely distrib-uted populations are roughly uniform, though continually changing. Clearly, in modern societies, any one person meets only a tiny proportion of the members of a natural language-speaking community. But there are tacit features of the language – such as the absence of certain sounds (words) that would indicate oddity in Baton Rouge buses – that are pioneered in narrowly bounded loca-tions – such as Baton Rouge. But these must be spread through face-to-face interaction because that is how language is learned. Therefore, the overall lan-guage must change through a process of diffusion *between* small and bounded groups. No doubt this process has been hugely speeded up by mass media such as radio and television, but it was happening before their advent: a dialect, or a certain frequency pattern of sounds and silences, develops 'here' but it dif-fuses to 'there' through the overlap of the speech of persons at the boundaries of local groups. The very uniformity of language over long social distances in modern societies shows how effective social diffusion is.[123] Though persistent differences in local accent also show that it is not a perfect mechanism, social diffusion is one way in which the local becomes general. Mostly, secondary localness works only with established and stable *ubiquitous* expertises – the things that belong in the topmost ovals of the fractal model of society: nearly everyone takes holidays and understands holidays so nearly everyone is in a position to offer warrants of reliability and create trust in their locality – every-one is an inadvertent ambassador.

The way trust first became non-local

There are a variety of ideas about how trust first extended beyond the local tribe or group. An early analyst of the problem was Herbert Spencer, whose answer was in terms of the interaction required to support the complexity of large societies – Spencer's view can be seen as a precursor of exchange theories and economic analyses which turn on individuals gaining benefit from interaction. The more well-known analyst is Spencer's successor, Emile Durkheim.[124] For Durkheim the tribe was held together by its 'mechanical solidarity', which

---

[123] This is not to say that local dialects and other patterns of speech do not survive too, but the uniformity of language over long distances is still remarkable.

[124] See Corning (1982) for a comparison of Spencer and Durkheim.

meant similarity of beliefs and activities among all the members, perhaps rein-forced by religious fervour.

But as societies evolved, the mechanism, though still supported by this early experience, became characterized by, Spencer-like, 'organic solidarity', the mutual cooperation required to support the division of labour; organic solidar-ity allowed for differences in skills and understandings within a much larger trusting society. Many theories of how larger societies hold together stem from Durkheim in one way or another, with the more economics-based theories finding a Spencer-like view adequate: the idea that market transactions require trust, so markets *make for* trust, since it is in every economic actor's interest for markets to function, are another approach. A problem with these theories is that, understood one way, they are teleological: the trust is 'caused' because it gives rise to good experiences or good outcomes. If this is the case, we need a mechanism to explain how the good outcome feeds back into the trust before it has happened. In evolutionary theory the mechanism is survival of the fit-test over many repeated tests of the same random changes, but we do not have enough repeated opportunities for this mechanism to work here.

The psychologist, Norenzayan, though giving almost no attention to Dur-kheim, takes religion as the key.[125] Gods prescribe ways of living and 'Big Gods', the gods of successful religions which spread widely, prescribe ways of living across large domains, enabling trust between fellow believers who accept the same God, even though they be strangers.

Networks

A different way of squaring the circle of trust and distance is with network-based theories. One influential theory is Granovetter's 'strength of weak ties'. Granovetter proposed that strong network relationships rapidly fold back to their origin in local groups whereas *weak ties* can convey information over large social distances in just a few long-distance jumps from group to group.[126] Note that the term 'weak' connotes two things: weak means 'structurally weak' because long-distance ties are likely to be singletons, whereas there will be many short-distance ties folding back from local groups creating, in network terminology, 'clusters'; but weak also means unpersuasive in the emotional sense because they will come from strangers rather than from friends and acquaintances and others more likely to resemble the recipient and maximiz-ing the effect of homophily and the like. The two kinds of meaning of 'weak' tend, therefore, to occur together in practice, and long-distance ties coming from strangers tend to be unpersuasive. Thus, though in structural terms weak ties seem to explain the spread of influence across long social distances, Gran-ovetter's idea seems more to redescribe the problem of remote trust rather than solve it, since we still have to ask why the distant strangers linked by weak ties

---

[125] Norenzayan (2013).
[126] Granovetter (1973).

would trust what travelled along them. Granovetter is well aware that trust is generated by the similarity – homophily – and shared presence that comes with co-location, and it is the contrast with this established idea that gives the weak-ties paper its initially striking, counter-commonsensical, impact. The resolution is surely that weak ties work well for the kind of *information* that everyone is ready to trust from the outset – depending only on transactional trust – but not for new *knowledge* of the kind that needs to be backed up by moral warrants if it is to be influential, especially if action based on it has a significant cost or significant risk (if it involves 'difficult and dangerous trust').

Recently, this point has been nicely argued by Centola who looks at what he calls 'complex contagions', transmitting the kind of thing that we associate with moral trust. Centola's argument, like ours, says that complex contagions involve risky investment of time and commitment. Centola shows with computer simulations and reports of existing empirical studies that the spread of new knowledge of this kind, and the behaviours that relate to them, require 'wide bridges' – many links reinforcing the same message – which, in regular social networks, are going to be of local origin; the kind of change in thinking or action that requires investment and moral commitment will not be transmitted through weak ties. To repeat, the only way that knowledge of this more complex and risky kind will be taken sufficiently seriously to be acted on is if its value is reinforced by many others, and normally many other similar people in a local group – a wide bridge.[127] In terms of the health analogy which Centola favours, 'simple' contagions like influenza or AIDS can be passed on by a single contact, whereas complex contagions require a number of contacts – hence the bridge has to be wide; Granovetter's weak-tie idea fits only simple contagions – or the spread of simple ideas.

Nevertheless, complex contagions and their equivalent can still spread through social networks in the way that natural language diffuses from locality to locality; this is social diffusion. Even local groups characterised by relatively closed networks may have influence across large social distance by social diffusion – one network influencing its neighbouring network, which influences another, and so on just as we have suggested for the way natural language diffuses across nations linking multiple, close-knit families which are the sites of linguistic socialization. We will deal with network theories at much greater length in Chapter 8.

Agents

Another approach is to posit that our trust in distant locations is provided by agents who report or work for us in a formal way.[128] But as Shapiro points out, this simply moves the problem of trust back a step: how can one be sure that

---

[127] Centola (2018, see, e.g., p. 7).
[128] Shapiro (1987, 2005).

one can trust the agents since what they do is likely to be opaque to us? One solution, of course, is to employ agents with whom we have personal relations, but this just returns the problem to the starting point. Another is to create additional agencies with the duties of overseeing the work of primary agents. These might be organised professions who certify and control solicitors, surveyors, and the like, or certified accountancy firms, or the rating agencies. Once more, as Shapiro points out, this pushes the problem back yet another step – who guards the guardians? Banking scandals and crises reveal the problem to which we have already alluded: the second-order guardians depend for fees on those they are certifying and are in competition with one another for business; this creates enormous pressure on them to provide positive reviews.

The possible ways of developing long-distance trust depending solely on remote communication are set out in Table 6.1 with the first row having been discussed in earlier chapters.

Table 6.1: How trust might be developed over long social distances.

HOW TRUST MIGHT BE DEVELOPED OVER LONG SOCIAL DISTANCES		
1	Distributed Ledger	Technological developments, such as block chain or multiple internet reviews of hotels and so on, create trust where the desired end is well defined and short-term	
2	Local Ambassadors	Where the skill or expertise is ubiquitous, a host of others can provide reliable experience, thus converting the problem of remote trust into a problem of local trust in acquaintances and neighbours while informative sounds and silences are found in local languages	
3	Social Diffusion	Social diffusion allows local languages and trust to spread reliably throughout modern societies	
4	*Market Exchange*	*Economic markets and social exchange in general create a need for trustworthiness*	*THESE TWO ROWS ARE TELEOLOGICAL UNLESS*
5	*Organic Solidarity*	*Division of labour creates a need for trustworthiness*	*A CAUSAL MECHANISM IS SPECIFIED*
6	Big Gods	'Big gods' extend Durkheimian local religious solidarity to wider communities who can be trusted to adhere to the same religion-based prescriptions on moral action	
7	Weak Ties	Local networks of personal relationships spread their influence widely through weak ties	
8	Supervising Agencies	A hierarchy of agents and guardians provide trustworthy services, though these may need bolstering with personal relations	

Conclusion to Chapter 6

What we think we have accomplished in this chapter is to explain the existence of distant trust – that is, to explain the growth of mass societies – in ways that are not incompatible with our argument that the establishment of 'difficult and dangerous trust' and therefore 'difficult and dangerous truth' is a matter of a small groups and face-to-face communication. Table 6.1 includes all the possibilities we have discussed in the chapter, but we have also shown that that several of these possibilities do not resolve the problem in a satisfactory way: some are tautological, some are limited to transactional trust and some still turn on the local. Short-term transactional trust, which is not hard to understand. is not our concern. We have argued that many instances of long-distance difficult and dangerous trust are really instances of local trust found in the reports of friends and acquaintances who have pioneered the expeditions, like early explorers to distant lands, our trust being based on what they report back to their local environment (line 2). And we have argued that much of what we securely know of distant locations is found in the local bath of words and silences absorbed in face-to-face interaction. What we have tried to do in this chapter is to show that the existence of long-distance trust relations, even where those relations are not transactional, does not vitiate our argument that societies which value sound knowledge must understand that, generally, and for the foreseeable future, this must be based on face-to-face relationships; they must not be tempted to replace the face to face with internet-mediated relationships.

CHAPTER 7

The 'Stickiness' of Face-to-Face Communication: Some Case Studies

There is lots of anecdotal evidence concerning the importance of face-to-face communication in resolving disagreements and moving them forward in a productive direction. Often, we just know that the best hope of resolving some deeply held disagreement is going to be a matter of 'talking it through' in some kind of socially engaged circumstance. We just know this without reflecting on it: it is part of the taken for granted in our social existence, at least in cultures that the authors are familiar with at first hand. Thus, in the mid-1990s, at the height of the 'science wars', Collins arranged a meeting between the two sides and, without consciously knowing why, rented a motorboat and driver to cruise the 10-strong party from both sides of the Atlantic, around the River Hamble and Southampton Water for a day with snacks and beer. It 'worked' and led to a book which the editors still believe was an important contributor to the Science War armistice.[129] Another anecdotal account involving water both as topic of the dispute and as the means of transport that settled it can be found in John Fleck's 2016 book *Water is for Fighting Over*. In this case, key decision-makers took a trip on the Colorado River and which, he claims, led to a 'breakthrough' (p. 162) that '[changed] the politics of Colorado River water management' through fostering a 'fragile bond ... that would strengthen during the coming years into collaboration' (p. 53). In Fleck's analysis – and the recollection of the decision-makers involved – taking a trip together created not only a 'change in attitude ... that rippled out through the Colorado River Basin problem solving for years to come' (p. 56), but also was the genesis of a working group that would go on to prove pivotal in collaborative management.

Now we continue Part II with some brief case studies which bring out the 'stickiness' of face-to-face communication. The question addressed is: What has

---

[129] Labinger and Collins (2001).

How to cite this book chapter:
Collins, H., et al. 2022. *The Face-to-Face Principle: Science, Trust, Democracy and the Internet*. Pp. 115–139. Cardiff: Cardiff University Press. DOI: https://doi.org/10.18573 /book7.h. Licence: CC-BY-NC-ND 4.0

made face-to-face communication so hard to abandon that even in domains where the remote is central to the activity, the face to face still flourishes? That the face to face still flourishes is telling us that some explanation for its importance, such as the ones we have worked out, are needed because it will not go away in spite of the explosion of new technologies. We allow ourselves to include various elements of intrinsically interesting material in each study, and readers who want to continue to follow the argument and the evidence without feeling the need to be persuaded by accounts of how the face to face plays out in diverse corners of society can skip these case studies or return to them later. The four main ones comprise: Linux, open source, software developers; a very large, multiple-site, international retail store; the management of emergency forest fire relief; and the shift to remote contact between university staff and students during the Covid pandemic.

Before embarking on these four main studies, we add a couple of paragraph-long anecdote-style sketches of the problem which did not invite writing up at greater length. First, Alun Preece is Professor of Intelligent Systems at Cardiff.[130] He has taken part continuously in US/UK International Technology Alliance projects since 2006, funded by military sources. These have involved industry researchers, academic researchers, government researchers and representatives of the end-users of the research (e.g., military advisors). They focus on technologies to support information exchange and decision-making in multi-partner coalition operations often involving AI and machine learning, and human-machine collaboration. But while this may appear to be an ideal setting for the functioning of remote communication alone, and the collaboration does make huge use of remote communication, it is still thought essential to embed face-to-face communication in the work. Thus, each funded researcher is provided around $10K per year for travel; this is considered necessary to reach consensus and manage plurality of approach. There is also an 'open campus' in Maryland which hosts visitors from home and abroad. Because of the intense technology focus of this group and its facility with intelligent machines and international collaboration, this is, again, an excellent illustration of the 'stickiness' of face-to-face interaction in technological circumstances which might seem ideal for its abandonment.

Second, Collins undertook a survey and completed a draft paper on attempts to introduce an email consultation system into a big, UK-based general practitioner practice with a view to freeing up the hugely overburdened face-to-face system. The attempt turned out to be a failure, encouraging more demand rather than less. But most of the reasons for this were technical or similar: a medical consultation often requires direct examination of the patient, or the careful taking of a history, so initial description of symptoms over email resolves nothing; distant diagnosis and treatment of children is far too risky;

---

[130] Preece was initially to have played a larger part in this project and his report of this briefly mentioned case was important in shaping the thinking.

incomplete descriptions require follow-up phone calls and usually lead to face-to-face consultations, so a single encounter becomes three encounters even though two of them are remote; patients use the email system as a way of circumventing a jammed appointments system which is limiting felt demand, thus resulting in increased demand! Bureaucratically insensitive use of the email system by doctors could lead to decreased demand but this was generally thought to lead to poor medical work. But none of these problems bear on our main thesis and problems of this kind are not represented in Table 1.1 because they are specific to medical consultations, rather than related to generic features of the local and the remote. For this project the two interesting problems – the ones that do relate to Table 1.1 – were reported less often in the survey of doctors, but the reports were still significant: some doctors reported that they occasionally suspected something was being hidden and a face-to-face consultation was a way to establish rapport and trust; some doctors felt only face-to-face consultations could lead their patients to trust them rather than what they heard on the internet or social media.

The main part of this chapter now begins with a case study which fulfils the role of an especially 'hard case' – a case which, if the need for face-to-face communication is experienced 'here', it is relatively safe to say will be experienced everywhere in the foreseeable world, because 'here' is the sort of place where we would least expect it to play a role, since remote communication is at the very heart of the activity. In other words, this case is a strong test of the Face-to-Face Principle.

The case of Linux kernel developers

The work of the community of Linux kernel developers is conducted *almost* entirely online and they are outstandingly successful in working and communicating remotely.[131] The substance they deal with, like that dealt with by the group briefly described above, in which Preece participates, is computer programs which, unlike words and their meanings, can be transmitted remotely without loss; this gives all such groups a tremendous advantage because at least a subset of claims can be immediately tested by replication without the usual doubts associated with the tester's competence and which give rise to the experimenter's regress. And, of course, workers in such areas start out as among the most skilful and technologically savvy users of remote communication technology that there are. If face-to-face communication plays an integral role in

---

[131] This description was first drafted by John McLevey, and the description of the functioning of conferences is based on McLevey's fieldwork observations at Linux and other open source software conferences between 2016 and 2019.

the community even here, it will support our argument about the irreplaceability of at least some face-to-face communication still further.

Linux is a computer operating system that plays an influential role in nearly everyone's life, usually indirectly, and usually without their knowledge. Although it is not popular in the personal computer market, it is by far the most important and pervasive system in the world of servers, which are vital in the current age of cloud computing, automation, computational science, complex global travel systems, smart devices, and so on. The Linux kernel is used in air traffic control and railway systems; nuclear submarines and defence systems; Android phones; most TVs; DNA sequencing software; the International Space Station and the Large Hadron Collider; the cloud computing servers used by Google, Amazon, Facebook, and so on; 95% of the top 1 million web domains; 80% of financial transactions; and nearly all of the world's supercomputers.[132] The 'Linux kernel' is the lowest-level code, which means it is the foundation of all the other routines, within the Linux operating system. As of 31 December 2018, the kernel alone was composed of over 26 million lines of code resulting from around 50 million proposed new lines being edited down by about half. This work was the result of contributions by thousands of developers from hundreds of organizations located across the world. Linux is the largest collaborative software project in the world.

The Linux kernel is 'open source', which means that all of the source code is freely available online and it is continually building as members of a worldwide network add new developments on a voluntary basis. A new version of the Linux kernel is released every two months. Each two-month development cycle represents the work of a relatively stable body of about 1,500 developers. Many kernel developers earn their living from one of around 400 organizations – they might be private sector organizations or non-profit, organizations and so on – that have a stake in the Linux kernel. The 400 include many large technology companies such as IBM and Google. Because of the unusual combination of voluntary contributions from the private sector with a freely available product, the collaboration has already been the subject of attention from researchers in the social sciences concerned with understanding what makes people contribute in the absence of direct reward. Here we look for the first time at the roles and relative balance of face-to-face and remote communication.

The kernel community coordinates their efforts and reduces the complexity of working at such a large scale by enforcing the use of formal processes and specialized tools. In fact, the official documentation that supports the development process makes a point of emphasizing that 'much developer frustration comes from a lack of understanding of this process, or efforts to circumvent it'.[133] In

---

[132] These and other uses are reported on the Linux Foundation website.

[133] The kernel development process is described at length in the official project documentation, which can be found here: https://www.kernel.org/doc/html/v4.15/process/2.Process.html.

brief, the development process consists of developers submitting code to official subsystem 'maintainers', who review the code before making a decision about whether it should be accepted or rejected. If it is accepted, it is tested extensively and submitted further up the chain to Linux's founder, Linus Torvalds. Torvalds has the formal, final say about what does and does not get included in the mainline kernel code, although he has been less involved in the project in recent years. According to the project documentation, there are about 1,000 changes a day. While the chain of submitting code from developers, to maintainers, to Linus Torvalds, dramatically reduces the complexity of making decisions in a project the size of the Linux kernel, in practice it is not possible for Torvalds or other high-level developers to personally access every piece of code that is submitted. As described in the official documentation:

> There is exactly one person who can merge patches into the mainline kernel repository: Linus Torvalds. But, of the over 9,500 patches which went into the 2.6.38 kernel, only 112 (around 1.3%) were directly chosen by Linus himself. The kernel project has long since grown to a size where no single developer could possibly inspect and select every patch unassisted. The way the kernel developers have addressed this growth is through the use of a lieutenant system *built around a chain of trust.* (emphasis added)[134]

The process of developing, reviewing, and making decisions about code happens almost exclusively in 'public' mailing lists and in the logs written by developers using the version control system known as 'git' (which was also created by Torvalds specifically for kernel development). This means that given sufficient technical understanding, the process can be inspected online, and this has been important to the analysis carried out here. Although we are primarily reporting on observational data in this study, our analysis is informed by extensive quantitative analysis of archival data from mailing lists and version control logs (specifically the structure and evolution of communication networks and collaboration networks in Linux development).

Face to face in kernel development

As can be seen, the power of remote communication in the case of Linux is enormous. However, there are things that remote communication does not do even in this case. It is a truism, of course, that no developer ever wrote their first script for the Linux kernel in complete personal isolation from the community

---

[134] As explained in 'How patches get into the kernel'. https://www.kernel.org /doc/html/latest/process/2.Process.html?highlight=merge%20 window#how-patches-get-into-the-kernel.

of software engineers. To function capably, remote-to-remote, in a new programming domain, developers need a solid foundation of skills, knowledge and experience that have already been acquired. For many developers (including Linux developers), these skills are acquired in classrooms, small startups, large commercial technology companies or makerspaces; these are some of the many ways in which acquiring the knowledge and skills required for making meaningful contributions to the Linux kernel are honed in face-to-face contexts. In developing many of their relevant skills in satellite communities such as the corporations they work for, face-to-face interaction will be part of their planning, development and testing.[135] As a consequence, the kernel community *does not need to be responsible for transferring core technical competencies*, many of which depend on the acquisition of the tacit, and *would* therefore require extensive face-to-face interaction– the Face-to-Face Principle at work!

The relative lack of importance of the transfer of foundational skills and competencies at Linux kernel conferences can be seen when compared with other types of open source and technical conferences. Many open source conferences outside of the Linux community generally have extensive workshops and session tracks oriented towards less experienced developers. For example, in a series of conferences focussed on the open source programming language Python (PyCon), talks on technical topics like Big-O notation and 'under the hood' improvements to speed and efficiency were held alongside talks intended for more novice audiences, such as motivating and explaining best practices for writing code that can be easily understood by other humans (PEP8 for Python) and when, why and how to refactor code. In addition to the conference talks themselves, these and similar conferences include extensive workshops and other opportunities for novice programmers or experienced programmers who are starting to contribute to or manage open source projects for the first time (e.g. on how to understand and navigate the differences between the many different free/open source software licences).

In addition to formal talks and workshops, these open source conferences have many informal learning opportunities. It is part of their culture; they are as much about learning, collaborative problem-solving and celebrating shared interests as they are about networking and rapid sharing of knowledge. For example, at many open source conferences novice and experienced programmers fill rooms and hallways working collaboratively on 'sprints' to complete small projects during the conference itself. At one Python conference in 2018, McLevey (who was a participant observer at the conference) was approached by a young novice developer who was working with a few other novices on a sprint to make a text-based adventure game. They were struggling with how to simplify the complex nested conditional logic of their game and but did

---

[135] This can give rise to tension since it can be in the interests of a firm to encourage the development of the kernel in a direction which favours their interests more than those of the community of users as a whole.

not know where to start. After a few minutes of conversation, several experienced developers who were working nearby joined the conversation, expressed enthusiasm for the idea of the game, supported the young novices in their first substantial coding project and offered their thoughts about how to most effectively redesign the game code.

The Linux kernel conferences also have novices in attendance, of course. And the talks and other conference events vary in their technical complexity and intended audiences. However, it is not necessarily the case that less technical talks are intended for novice programmers; they may instead be intended for people who manage kernel developers, or who want to contribute to the kernel community in ways other than actually writing code. Events that are explicitly oriented towards new contributors tend to focus more on networking than on learning core skills. For example, in 2016, McLevey spent a day at a LinuxCon observing events designed to connect new kernel developers with experienced kernel developers and maintainers. Conversations between the two levels of developers covered an extremely broad range of topics, but almost none of them seemed to discuss technical topics related to kernel development. Instead, new kernel developers seemed most interested in using the opportunity to meet and chat with high-profile developers and to emphasize their technical abilities by describing the types of projects they had undertaken in their career to date. In other words, at least at this particular LinuxCon, these and similar events were more about networking and social status than they were about learning; most of the time, both parties in a conversation were accomplished developers, but one was deeply embedded in the kernel community and the other was new to it. In sum, these meetings were more about forming links between peripheral and core developers in a massive collaboration network with a core-periphery structure than they were about novice developers learning from accomplished developers.

We are *not* arguing that transparent networking and displays of status always happen at LinuxCon and that they never happen at other conferences, open source or otherwise. Obviously, that is not the case. However, technical communities develop cultures just as other communities do, and a strong focus on acquiring and sharing *foundational* skills is much less salient in the culture at Linux conferences than it is at other open source conferences.[136] The key point we want to make here is that the kernel community does not need to focus on using limited face-to-face time at conferences fostering learning and professional development for novice programmers who *need* face-to-face interactions to acquire tacit knowledge and hone their foundational skills. The kernel community is a highly specialized technical community. Generally speaking,

---

[136] Note that is not to say that they do not occasionally provide *advanced* training, but even this is less common and because the developers are all experienced, learning happens much faster and more effectively than it does for novices, and so it is a much less visible part of the conference culture.

newcomers are new to the Linux *community*, not to software engineering and certainly not to computer programming more generally. The kernel community, then, is a kind of second-order technical community that sits on top of more primarily technical communities. It benefits from an entire ecosystem of learning opportunities – including other open source conferences – where novice developers hone their skills. This enables them to use their face-to-face interactions towards different ends, such as planning, strategy, building emotional energy and maintaining interpersonal relationships that hold the network together in the absence of formal organizational boundaries and structures.

In spite of the huge advantages it has as a social community when it comes to availing itself of remote technologies and procedures and peripheral organizations to manage basic socialization into computer programming skills, face-to-face meetings are still a central part of the business of Linux kernel development activities. They are not a large part in terms of time spent, but they are tremendously important in terms of role in the community. One way in which face-to-face communication matters in the kernel community is through the process of transforming 'relational tacit knowledge' into explicit knowledge. Beyond that, the role of face-to-face communication differs for the relatively small subset of highly active developers and maintainers ('core' contributors) and for the rest of the kernel developer community. We discuss these processes below.

Relational tacit knowledge in the Linux kernel community

One complicating feature of this rapidly developing community is that much of the shared tacit knowledge is 'relational tacit knowledge'.[137] Relational tacit knowledge, unlike 'CTK', can be explicated. It is part of our tacit knowledge because it is impossible to explicate all of it, only a 'moving window' – tacit knowledge comes in at the 'left', as it were, becomes explicated and goes out at the 'right' in explicit form while new relational tacit knowledge is always appearing the left as the world evolves. Because this is a rapidly developing domain, what were once aspects of being a kernel developer that depended on face-to-face interaction are now explicitly documented in one place or another. For example, in 2014, an especially important subsystem maintainer Greg Kroah-Hartman gave a talk to Linaro employees called 'Why I don't want your code: Linux kernel maintainers, why are they so grumpy?' that explicitly laid out the realities of work as a subsystem maintainer, revealing experiences that make it all too clear why seemingly small things – for example converting tabs to quadruple spaces, including privacy boiler plate in email messages, using an incorrect coding style, submitting multiple patches without specifying their build order, and so on – can determine whether or not your code is

---

[137] Collins (2010).

accepted by a maintainer. This is interesting in part because the list of potential violations is very long and, in theory, *there is no limit* to the number of small things that might lead to rejection. Kroah-Hartman's list is a partial accounting of examples from a two-week window, and in articulating them he was making relational tacit knowledge about the realities of being a maintainer explicit. But there are *many more* things coming in the left side that remain tacit.[138]

Thus, though these aspects of the once tacit can now be handled remotely, new examples are always appearing. Relational tacit knowledge can give the impression that because there is a process whereby some of it transitions to explicit knowledge, everything tacit can be made explicit with enough time and effort. But this is wrong both because there are other kinds of tacit knowledge and because relational tacit knowledge itself is being continually renewed.

The difference between experienced and novice developers

As previously mentioned, there are some key differences in the role that face-to-face communication plays for the subset of core developers and maintainers and the rest of the kernel developer community. For example, the core developers have a lot of remote interaction with peripheral members, but they *also* have extensive face-to-face interaction with other core developers in less public settings. Perhaps the most important example of this is the annual Linux kernel maintainer summit meeting, where a small number of kernel maintainers and other elite kernel developers discuss development processes, including analyses of previous development cycles and planning for future development cycles. These are small, invitation-only sessions that explicitly prioritize agenda items that are best reserved for face-to-face exchanges. In an email kicking off planning for the 2019 kernel maintainer summit in Lisbon, a member of the programme committee wrote:[139]

> Linus has a generated a list of 18 people to use as a core list. The programme committee will pick at least ten people from that list, and then use the rest of Linus's list as a starting point of people to be considered. People who suggest topics that should be discussed on the Maintainer's summit will also be added to the list for consideration. To make topic suggestions for the Maintainer's Summit, please send e-mail to [REMOVED] list with a subject prefix of [MAINTAINERS SUMMIT].

---

[138] Ironically, Kroah-Hartman's talk transforming some relational tacit knowledge about maintaining into explicit knowledge for developers was delivered face-to-face to a small room of developers and managers, but is publicly available on YouTube for remote viewing.

[139] The link to the email being quoted here is: https://lwn.net/Articles/788378/.

The other job of the programme committee will be to organize the programme for the Kernel Summit. The goal of the Kernel Summit track will be to provide a forum to discuss specific technical issues that would be *easier to resolve in person than over e-mail.*

The email goes on to explain that some 'informational' topics might be considered for discussion if they serve the interest of the developer community, *including advanced training.*

The programme committee will also consider 'information sharing' topics if they are clearly of interest to the wider development community (i.e. advanced training in topics that would be useful to kernel developers).

Finally, the same email ends by explicitly referencing the importance of spontaneous hallway conversations that we have been arguing are so central to scientific and technical communities.

We will be reserving roughly half of the Kernel Summit slots for last-minute discussions that will be scheduled during the week of Plumber's, in an 'unconference style'. *This allows ideas that come up in hallway discussions, and in the LPC miniconferences, to be given scheduled, dedicated times for discussion.*

Of course, the elite developers who attend the kernel maintainer summit and other core developers who are essential to developing and maintaining Linux, do not only interact with one another, they interact *extensively* with developers who are much more peripheral to the community. At Linux conferences, these interactions include all the usual types of activities at tech conferences, such as eating, drinking, collaborative work sessions, late-night dance parties with DJs, etc. Core and peripheral developers both participate in these events together. These events build up emotional energy through 'interaction ritual chains' that sustain remote development efforts.[140] As one interviewee put it, tongue in cheek, for the typical Linux kernel developer, this is enough face-to-face interaction to last quite a while.

Perhaps even more important than emotional energy, if not entirely separate, is the fact that core developers can use their face-to-face time developing relationships in ways the significance of which are indicated in Table 1.1 and (a) are difficult to manage remotely and which (b) are useful for working effectively with and getting the most from one's collaborators. Eating and drinking together, even once a year, increased the likelihood of delivery on commit-

---

[140] The phrase is due to Randall Collins (1998; 2004) drawing on the work of Erving Goffman (1967).

ments. It is likely that at least some developers do what they do in the Linux community because of the relationship and community building that happens in face-to-face settings. Mailing lists cannot do this for several reasons, one of which is that they have a reputation for being hostile environments; they do not paint a picture of a welcoming community where contributions are welcome from just anyone.

Linux kernel conferences help build emotional energy, they enable developers to cultivate relationships in ways that are difficult to do in remote settings, they enable core developers to have discussions about the future of the Linux kernel and to plan strategically in an ambience of trust. They even *plan* for spontaneous conversations and restrict face-to-face conversations to topics that are difficult to discuss over email. All these factors are important for the community *even though* they may not be necessary for your average Linux kernel developer to get code into an official release of the kernel. Any individual developer, given enough technical expertise, could rely on tacit knowledge acquired from other communities and explicit knowledge about the kernel development process to get code accepted into the kernel code base. But the community itself relies on extensive face-to-face interactions that are easily overlooked, given the tremendous success of remote communication in this case and the way that important elements of face-to-face learning are hidden away in satellite locations.

The case of Huge Stores

Big business might be thought to be wide open to the replacement of face-to-face communication with remote communication, since much of what it does depends on transactional trust. Therefore, the fact that face to face will not go away even here is worth pointing out. We have already made the point about the efficiency of moral trust as opposed to transactional trust even in the case of business. Here we illustrate this with a more extended example.

This material resulted from a serendipitous co-location of one of Collins's acquaintances in the gravitational wave domain and an important business-man, 'Top Guy', who had mentioned to Collins's acquaintance that top firms often relied on personal relationships.[141] Collins asked his acquaintance to follow up the remark and it is this that led to further discussion. The material is from an initial email from Top Guy to the acquaintance, on 8 January 2019, then a telephone interview between Collins and Top Guy. To repeat, it is important to contrast this with the purely economic model of business transactions – a model that would suggest that technical replacements for local trust ought to be readily available. First, the email comment:

---

[141] William D. Cohan's (2021) guest essay for the *New York Times* makes a similar point with respect to the world of investment banking.

Most large retailers have suppliers on the ground. For instance, I believe Unilever have >100 and Proctor & Gamble has >300 in the region of Huge Stores headquarters. The reason is to foster a long-term and collaborative relationship that can only happen F2F, Skype etc will not enable that, we are still physical creatures.

The telephone interview with Collins was then arranged.

Edited extracts from telephone interview, UK/US 15 January 2019

Huge Stores = *a big American retailer with multiple stores on the scale of Sears, Walmart, Tesco or Asda*
Storeville = *a smallish provincial town in which Huge Stores headquarters are located*
Big Firm = *a FTSE 100 company supplying Huge Stores*
Top Guy = *businessman responsible for a major stream in the supply chain from Big Firm to Huge Stores; he moved to and lived in the region of Huge Stores' headquarters for three years on his last posting*

T[op] G[uy]:   We are committed to work with all our partners in a very collaborative fashion. We find that face to face is not only the best way of instilling a relationship with our retail colleagues and partners ... we have big and long-standing businesses with these companies, but also it's the most efficient way, often, of getting stuff done. You know, you stand in a room with someone, you know, you can easily ... you can understand someone's body language very well, you know, there is something personal about – and builds trust ... especially if you've got more than a one-on-one – if you've got a broader group of people – you know, you obviously know who's speaking all the time. With Skype it is a little challenging to detect all the voices if you've got lots of people in the same room.

C[ollins]:   If I was a mainstream economist, I would be inclined to say that none of this would matter – all that matters is what I cover with my mathematical calculations about the price, and information exchange. So why on earth would people want to get to know each other?

TG: In circumstances where things are purely data-driven then I can understand why someone would say that. [In our industry], while we rely heavily on the data and trends ... essentially we are designing products for consumers ... so there's a lot of collaboration, thinking, brainstorming, exchange of ideas, needs to happen, based on what we think, our experiences, what we bring collectively to the table in terms of our diversity, that leads us to a good exchange of views and often to a better solution than merely looking at the data which can often mask the underlying.

C: I don't want to put words into your mouth, but is it important for you to trust your business partners?

TG: One hundred per cent. Without trust business becomes very difficult.

C: So could you give me some examples of how face to face develops trust and why trust is important?

TG: [Starting with the second one] Trust at the basic level is about whether what we agree is going to happen, whether that's an exchange of payment up until whether we agree to execute quite complex plans over thousands of stores involving thousands of employees or our staff. So there has to be high level of trust 'my word is my bond, kind of thing' – what we say we are going to do we actually do. Without that, things become incredibly inefficient so therefore trust is the currency of the business. I've met people in my time who have been buying [a certain consumer product] for decades and never signed a contract in their life; they just go on 'their word is their bond'. We are not quite at that level any more, but certainly trust is the lifeblood of a lot of what we do. And without trust, what happens is essentially, you have a lot of inefficiency. You have lots of checking, much more contract-bound, transactional relationships. It slows things down and makes everything more inefficient.

C: And how do you develop this trust with new people?

TG: People have different theories on this. In my personal experience trust is earned, first of all, it's never given. You earn trust in the most basic form by doing what you say you are going to do and delivering on whatever agreement you've made … over a period of time. That's the hardest kind of trust. The softer kind of trust is getting to know someone. To talk to people not just as business partners but also as people. That allows people to understand who they're talking to and where they're coming from. And also makes business a bit more fun as well – a little bit less automated and robotic. …

C: Do you remember inviting people to dinner when you first arrived in the location of Huge Stores?

TG: Different retailers manage things in different ways. Some will not allow gratuities of any shape or form, even going to someone's house is inappropriate. Under these circumstances, our primary connection is through F2F meetings, industry and community events.

C: … Who insisted on the policy that people should be local?

TG: Many retail partners encourage major suppliers to have offices in the locality to foster a sense of collaboration and joint business planning. They feel that together they can continue to drive cost down to the consumer and develop better products and drive growth for the benefit of both parties. Huge Stores felt that people being in and around Storeville would be the right thing. And if you look at the investment that Huge Stores has made in the community then you can see that not only do they still believe in that, they also believe that they want to continue to attract the best talent from their suppliers to Storeville, by making it an attractive place to live.

C: So you think I'd be safe to say that it's Huge Stores' initiative that began it all in their case?

TG: That's my understanding but you'd have to check the facts on that one … But I think you would be safe.

C: Do you know of any firms who do business with Huge Stores who don't have any representatives in the Storeville area?

TG: No, but there are also brokers who live there – there's a whole community of brokers – who will represent your company to Huge Stores – who live there on the ground and so what you'll find is that smaller suppliers will connect with that broker who will represent them to Huge Stores.

It is indicative of the importance of face-to-face communication that huge businesses like Huge Stores will encourage it when they have the power and resources to manage communication remotely and with what would be, under the economic model, greater efficiency with no travel costs or time-wasting personal meetings and, of course, faster transfer of information. Even in business like this, F2F is preferred.[142] This interview illustrates the point already made, that moral trust is far more efficient than transactional trust because there is no need for continual checking and because it works for things that cannot be checked.

The challenges of face-to-face communication in emergency management

The importance of face-to-face communication is also ingrained in the case of emergency management – where effective communication can mean the

---

[142] But see Appendix 4 for an alternative view emerging from the organization studies literature.

difference between life and death.[143] Its 'stickiness' is still more striking given that there are significant difficulties in this sector for arranging meetings: the geographical obstacles involved, different on every occasion, and the speed with which emergencies develop. Fortuitously, this is also a sector where less fraught post-event reviews are regularly performed. These reviews are a window into the communication challenges that emerged during the moments of crisis.

There is a difference between emergency *response* and emergency *management*. Emergency response includes the frontline services such as paramedics, firefighters, police officers, natural gas leak response teams, and so on. Here the importance of face-to-face communication is obvious. Paramedics, for instance, assess patients not just through their objective injuries, but through diagnostic questions and careful attention to body language and physical reactions. Their bodily presence helps to build rapport and trust and reassure the injured, calming them towards a point where they can offer coherent responses. Even firefighters, when dealing with fire not people, rely on physical sensations, like the feeling of the building underfoot, and smells or sounds, to assess the situation in a way that would not be possible using even video supported communication. This kind of imperative for co-location does not, of course, bear strongly on our main thesis about the relationship between face-to-face and remote communication which has to do with trust rather than the direct apprehension of symptoms or problems: these factors are, again, not of the kind represented in Table 1.1. Nevertheless, we have to understand them so that we know exactly what we are learning from cases like this.

Emergency *management* focuses on the preparation for, management of and recovery from a crisis. All three involve trust and reliance-building. Emergency managers work with community members to develop plans for how a team will respond during a disaster, requiring levels of trust and reliance that will withstand the psychological pressures of an emergency. They are also involved in the recovery phase, helping communities return to a sense of normalcy which can turn on the personal investment manifested in physical presence.

On the face of it, however, face-to-face communication by managers seems out of place *during* a crisis. Imagine a typical emergency response situation, such as a forest fire in Canada. The province of Alberta spans over 660,000 square kilometres – almost three times the size of the entire United Kingdom. Much of the province is also heavily forested, which creates the possibility of extreme wildfires. These fires can be frequent, massive, and very quick to start and grow. Emergency managers have to decide how, where and when to dispatch firefighters; how many responders ought to be sent to which fire with what resources; and they may need to respond to changing circumstances in real time, for example additional helicopters or air tankers might be sent. They must also coordinate with emergency managers in other organizations, such as the Royal Canadian Mounted Police, the Alberta Emergency Management

---

[143] This section was first drafted by Eric Kennedy and is based on his fieldwork.

Agency, and local city governments and responders, each of whom are based in different offices in different parts of the province.

This seems a near-impossible situation for face-to-face communication, both within individual agencies and between different organizations. The primary headquarters for Alberta Wildfire is located in Edmonton, the provincial capital, which can be hundreds of kilometres away from where a fire breaks out. Each other organization also has its own, separate headquarters, often including a large number of personnel and resources that cannot be left behind. For example, the Alberta Wildfire headquarters has a large meteorological team, a huge group of highly trained support staff, extensive computing systems, media briefing facilities, and an operational room with large screens and maps to maintain situational awareness. Moreover, when local command posts are established for emergency operations, they need to be located at a safe distance from the event itself – quite a challenge, given that aggressive fires can move dozens of kilometres each day. And, even if the decision was made to co-locate on-scene with the emergency itself, it would take hours or days for everyone involved to travel and set up such infrastructure.

Emergency management organizations are, then, faced with competing face-to-face communication priorities. At one extreme (see Table 7. 1), groups could choose to retain maximum face-to-face communication within their own organization by keeping personnel together in the default location. But this would reduce communication with other organizations and proximity to the incident itself. At the other extreme, one could physically co-locate as much as possible with the incident but would be faced with the challenge of perpetual relocations due to changing conditions, the difficulty of physically co-locating simultaneously with several concurrent incidents, and the time lag involved in establishing this co-located setting.

As we will see, most forms of emergency management tend to prioritize in-person communication for inter-organizational communication (the middle columns of Table 7.1), while spanning geographical distances represented by the grey columns with remote communications supported by trust and reliance established in earlier, non-emergency, face-to-face meetings.

An example of a choice

The wildfire that affected the City of Fort McMurray in May 2016 illustrates the kind of choice that has to be made. The fire destroyed over 3,000 buildings and caused almost 10 billion US dollars in damage when it rapidly burned through the city. While only two residents were killed – victims of a car accident likely related to fatigue from the evacuation – post-incident investigation suggested the incident management was not as effective as it could have been because of communication challenges. In particular, multiple 'After Action Reviews' investigating the event pointed to a common problem: the lack of 'unified command' during the incident.

Table 7.1: Different ways of balancing face-to-face and remote imperatives.

Central Headquarters	Near Scene Command	On Scene-Personnel
computational resources, large team, proximity to team members' homes, proximity to government	established at a distance from the event for safety, but designed to provide integrative location for multiple agencies	able to engage directly with the incident, but vulnerable to incident / distractions, and too close for perspective

During an emergency, common practice is to adopt a standardized method of interactions. In the Canadian context, this mechanism is called the 'Incident Command System' (ICS). It lays out specific roles that need to be filled (e.g. incident commander, finance officer, logistic officer, liaison officer and so on), as well as pre-defined job duties and principles for incident organization (for example, to create action plans for a specific period of time, setting goals and tasks, and revisiting and updating these plans on a fixed schedule). The ICS framework offers clarity during stressful, complex events: certain people are given certain roles, and certain roles are responsible for certain tasks. There is a clear hierarchy of who reports to who, and a fixed structure wherein all parties are kept appraised of happenings in the field. This formality is, in many ways, the equivalent of the use of checklists in aviation: it helps to ensure that no steps are missed, that responsibility is clearly demarcated between the parties and that a person responsible can make a quick decision if needed (rather than being unsure of the chain of command or pursuing unanimity).

Importantly, ICS is established within a physical command post located near the scene of the event. While these posts (and their larger, more formal equivalents, 'Emergency Operations Centers') can take on many different layouts and forms, they share a common principle: those holding a role within ICS are physically brought into the same location. This is critical because the formality of the system is insufficient for effective communication. Instead, incident command staff need logistical benefits that arise from face-to-face communication: rapid turn-taking; access and rapid interaction with shared resources (such as the ability to point to or draw on a map); and the investment of being entirely focussed and present in their assigned ICS role (not attempting to multitask with other roles, or being distracted by inputs from a different environment).

It is unsurprising that communications can break down when this physical co-location is not achieved. In the case of the wildfire in Fort McMurray, multiple parallel incident command posts were established, each independent of the other. The Regional Municipality of Wood Buffalo (the local government) activated its Emergency Operation Centre, which was joined by the Alberta Emergency Management Agency. Another incident command post was established within the city itself, while Alberta Wildfire established yet another command post for firefighting operations. The effects of this physical split – and the false illusion of effective communication that resulted – were striking.

According to an 'After Action Report' prepared by MNP in 2017:

> Various interviewees indicated that they 'were at the ICP' or talking to the appropriate command structure but, in fact, they were in different locations talking to different people. Clearly, communications among these structures was difficult and added to the complexity of gaining one clear operational plan for dealing with the wildfire … Those involved in wildfire operations were obviously disconnected from each other in many circumstances. Things changed rapidly over the duration of the Horse River wildfire, and the organization struggled to keep pace with changing developments. (p. 40)

Practical problems arose because communication was remote and between multiple venues. Industry representatives were largely left out of communication loops despite their need to plan and respond, illustrating the way that 'inefficiencies … created tensions and additional work among these partners' (p. 41). One moment illustrates the problem in a striking way:

> [By] late morning on May 3, the [the Operations Chief at the Alberta Wildfire Command Post] realized the wildfire was beyond the ability of firefighting resources and would run into Fort McMurray that afternoon. On the municipal side of the incident, another Operations Chief was deployed to the REOC [Regional Emergency Operations Centre]. Rather than learning about the wildfire's imminent incursion into Fort McMurray through the ICS structure, the [Regional] Operations Chief *discovered the wildfire was in the community through public reports over social media*. (p. 41, emphasis added).[144]

To summarize thus far, emergency managers are forced to deal with rapidly evolving, highly complex situations. The established best practice for doing so is through establishing a clear, hierarchical command structure for the incident. This involves specific roles, pre-determined lines of accountability and, critically, a centralized, person-to-person facility where these interactions can occur. The use of face-to-face communication increases the volume of information that can be exchanged, makes it easier to enlist the right personnel for a conversation, making sure that important contributors are not left out of the loop, and explicitly avoids back-channel and side-channel communications that can lead to crossed wires. When personnel are not physically brought together in a single venue, there is increased risk of miscommunication, sometimes with catastrophic results (like, as an emergency manager, only learning the wildfire has arrived in your community via social media).[144]

---

[144] Illustrating, by the way, one of the *advantages* of social media's openness!

It is worth making one more note given the earlier discussion of the geo-graphical impossibility of total face-to-face communication. Because this co-location of relevant parties is so critical, gaps open between the centre and outer columns (on scene and organizational headquarters) in Table 7.1. In other words, in establishing unified command, personnel must leave their headquarters and arrive at this command post, and the command post cannot be physically collocated with the incident itself.

The way these gaps are most productively managed is through relationships grounded in face-to-face communication. For instance, given that emergency managers cannot be on scene with responders, it is productive to ensure that at least some managers have lived experience on the ground and maintain active, face-to-face relationships with responders. Their initial experience provides, through face-to-face enculturation, the tacit knowledge, shared assumptions and perspectives required to improve the effectiveness of their remote com-munication. For instance, by having experienced fire firsthand, by knowing the language and perspectives of firefighters, and by having developed carefully calibrated reliance on frontline responders, the managers are better able to communicate effectively with their colleagues in the field. And ongoing encul-turation through face-to-face meetings (for example, shared training events, being part of the same social networks, deploying to out-of-jurisdiction mutual aid activities together, and working together in incident command posts) ena-bles them to maintain trust and reliance in current personnel.

Likewise, typically on-scene and near-scene command post roles (the centre and right groups in Table 7.1) are staffed by those who normally work at the headquarters. The history of day-to-day and face-to-face interactions are what allow them to work for short, intense bursts at a distance. For instance, when Kennedy observed the group that coordinates cross-jurisdictional exchange for wildland fire (the Canadian Interagency Forest Fire Centre) during the busy 2017 season in British Columbia, perhaps their most critical move was to send a representative from the main office to the unified command post in that province. Rather than trying to ascertain critical details from personnel with whom they had no significant relationships, the use of a core staff mem-ber created a face-to-face bridge that could span the geographical separation. Communication was effective between the remote staff member and the home office precisely because of the trust and shared knowledge built through years of face-to-face interaction.

In summary, then, the field of emergency management shows us cases where face-to-face communication is critical, despite being logistically diffi-cult. Failures to quickly and effectively establish face-to-face communication can directly cause operational failures in emergency response. Furthermore, while remote communication technologies offer distinct advantages (e.g. rapid deployment, communication that can span vast distances, and even simple paper-trails from email chains), distances are most effectively spanned through relationships and trust that is initially built face to face. The advantages of face to face range from the logistical (e.g. faster turn-taking) to the profound (e.g. trust and reliance), but are essential to effective emergency operations. In other words, even as remote communication technologies expand, their role is likely to be complementing, augmenting and enhancing communication that is ultimately rooted in the face to face.

Capturing university lectures for remote consumption

One example of the pressure to exchange the face to face for the remote is the move toward more distant forms of learning at university level.[145] The economic and logistic efficiency of such a move is obvious so long as university education is seen as transmission of information rather than enculturation into a body of ideas and new ways of thinking and knowing. But the use of lecture capture technology can also be justified by its advantages in respect of students with additional learning needs (ALN) or those with health or caring responsibilities that make it hard for them to be in control of the timetable of their lives. Such students can still attend live lectures if they wish or when they can, but a recording of the performance may give them opportunities to catch up on what they missed at their own convenience and speed. This kind of argument for the benefits of lecture capture is hard to gainsay, irrespective of worries about intellectual property ownership, and it means we now have the material to examine how lecture capture works in practice and compare live lectures with recorded lectures.

Lecture capture usually means 'voice-over PowerPoint' – that is to say, the recording consists of the slides and any other material displayed to students and the accompanying talk from the lecturer. Where the room is equipped with a video camera, a second display shows the lecturer as they give the lecture, though this is tends to be a fixed focus shot of the lecturer behind the podium and many lecturers choose to disable this feature. In what follows, our focus is the choice between going to the traditional face-to-face lecture, watching the recording of that lecture in one form or another, or doing both; we are going to try to establish that a recording is not as faithful a reproduction of a live lecture as it seems.

Evaluation and use: what is already known

Lecture capture has been around for some time and there have been a number of attempts to work out its impact and efficacy, from which some common themes have emerged:[146]

---

[145] This section was first drafted by Robert Evans.

[146] See e.g. Karnad (2013), Witthaus and Robinson (2015) or Nordmann and McGeorge (2018). Allowing or facilitating the recording of teaching is often

Effect on attendance is none to small: students who choose not to attend do so for other reasons

Use of lecture capture is relatively focussed: students do not watch whole lectures, but focus on small segments

Use of lecture capture increases with proximity to assessment, particularly exams

Overall use is relatively small, with significant proportion of the cohort not using it at all

Students with ALN value it more that those without

These trends are consistent with an evaluation of the use of lecture recordings by students in the School of Social Sciences at Cardiff University (SOCSI) carried out during the 2017–18 academic year when lecture capture was being piloted for the first time. For example:

Usage was relatively low at approximately seven clips or views per student; to put this in context, most students will have at least 60 lectures in any academic year

Students tended to watch clips or sections of lectures rather than whole lectures

Use of recordings peaked as assessments approached, with students also saying they used them for revision and/or assessments

There was a difference between students with and without ALN, with the former seeing it as more important

Lecture capture and tacit knowledge: what is missing

Whilst there has been much said about lecture capture and its impact on student learning, surprisingly little turns on the issues that are central to this book – socialization and the transfer of tacit knowledge. This seems even more odd since such a concept of university education is enshrined in the UK's official

---

identified as a way in which universities promote inclusive teaching (see, e.g., Disabled Student Sector Leadership Group 2017). There is also a large literature evaluating the effectiveness of the lecture with Bligh's *What's the Use of Lectures*, first published in 1972, a widely recognised classic that is now in its fifth edition (Bligh 2000).

guidelines. The Quality Assurance Agency for Higher Education (QAA) demands that students graduating with upper second-class honours (a 2:1, the modal outcome and the one most students see as the bare minimum they hope to achieve) meet the following standards (amongst others):

> The student has demonstrated sophisticated breadth and depth of knowledge and understanding, showing a clear, critical insight.

> The student has demonstrated the ability to make coherent, substantiated arguments, as well as the ability to consider, critically evaluate and synthesise a range of views and information. They have demonstrated a thorough, perceptive and thoughtful interpretation of complex matters and ideas

> The student has performed practical tasks and/or processes autonomously, with accuracy and coordination.[147]

Crucially, these skills of critical insight, thoughtful interpretation and acting autonomously all require the student to learn 'how to go on' – that is how to apply an idea, technique or heuristic – independently and in a new context. What counts as 'critical', 'thoughtful' or 'accurate' can only be understood in the context of the discipline being studied and, as has been argued in many places, the grasp of appropriate responses in context is something that can only be acquired through substantial immersion and social interaction in that context – one of the greatest obstacles to the creation of intelligent machines.[148]

Within the traditional higher education system, there are roughly three modes of teaching:

> lectures, in which information or knowledge is transmitted to students via lecturers;

> seminars or practicals, in which students use or apply this knowledge in discussions, experiments or similar 'labwork';

> independent study, in which students work alone or with peers to either practise skills or extend their knowledge by reading more widely.

Within this tripartite division, much of the socialization will take place in seminars and supervised labwork. Here they are immersed in the bath of words,

---

[147] Source: QAA Qualification and Credit Frameworks, Annex D (https:// www.qaa.ac.uk/docs/qaa/quality-code/annex-d-outcome-classification -descriptions-for-fheq-level-6-and-fqheis-level-10-degrees.pdf? sfvrsn=824c981_10).

[148] For example, see Collins (2018).

learn domain discrimination, experience immediate influence on interpretation, take advantage of body language modifying meaning, engage in safe adversarial dialogue, benefit from efficiency in conversational turn-taking, and offer and receive expressed commitments and injections of energy signifying the importance of some things and not others without it having to be made explicit – these being listed in Table 1.1 of this book.

In contrast, lectures seem much more like a remote form of communication. Whilst potentially interruptible (e.g. if a student were to raise their hand and ask a question) they are often more like a broadcast talk in which a single figure renders some body of knowledge explicit either by speaking or demonstrating its content to a largely passive audience (see Table 4.1). Indeed, given this, it is hardly surprising that lecture capture, and its extension into Massive Open Online Courses (MOOCs), seems like a natural development: if the only thing going on in the lecture theatre is the transmission of explicit knowledge to the audience, there is little to lose by viewing the recording alone and much for institutions to gain by, for example, making the intellectual work of their employees more widely available.

This, in turn, explains the enthusiasm, evident at Cardiff University, for re-purposing lecture capture to support 'flipped classrooms' and other forms of blended learning, despite the fact that the extra emphasis on small groups results in increased rather than decreased logistic demands. In the flipped classroom, the above critique of the lecture is accepted, as is the threat it poses to traditional universities from online courses that offer the 'same' product at a lower price and in a more convenient format. In response, academics and universities who advocate this approach seek to reduce the role of lectures in university teaching, arguing that this kind of information transmission no longer requires gathering students together in a single place and can be accomplished more efficiently with pre-recorded (mini) lectures that students access in preparation for less structured 'workshops' or other, more interactive, forms of teaching. In essence, the idea is that a university should provide access to academic staff, and flipped classrooms enable this by replacing the timetable slot occupied by the lecture with something a bit closer to the seminar. This response to the new technological possibilities once more illustrates the 'stickiness' of the face to face.

Tacit knowledge and the collective experience

But there is still something missing from this line of argument. It accepts the premise that the lecture can be reduced to the transmission of explicit knowledge from a single source to multiple recipients but that may not be true. Perhaps those students who continue to turn up to lectures, and 'fail' to take advantage of recorded lectures except in the most minimal way – a few examining short clips at most – know something that we have not yet considered here. Returning to Table 1.1, we can identify the following benefits of being physically co-present in lectures, some of which are shared by workshops and seminars but none of which are provided by the lecture recording alone:

1. **Domain discrimination**: Whilst this is likely to be less intense than the experience generated during seminars or workshops, lectures do enable students to see and, to some extent, interact with lecturing staff. The fact that the lecturer arrives early or stays behind to answer questions is not visible on the recording but can be seen by those in the room, which might, in turn, help create an impression of that person as someone to be trusted. Likewise, students can see each other and monitor their reactions to content and ideas. Seeing that others look confused when you feel the same will be reassuring for some, whilst pre- and post-lecture discussions amongst peers may also help to clarify and share ideas. To the extent that watching a lecture recording remains a solitary experience, these benefits of collective learning – and the deeper processing of lecture content it promotes – will not be available to viewers.

2. **Immediate influence on interpretation and body language modifying meaning**: This may be a particular problem with lecture capture that is available as 'voice-over PowerPoint' as any signs, gestures or other movements designed to give emphasis or illustrate a point will not be visible to the remote viewer. In this sense, recordings, like other forms of explicit knowledge, are always to some extent decontextualized and incomplete. It is also true that, at least for some lecturers, the presence of the audience alters the way the lecture is presented and, to the extent that they are not present, so the body language of those watching remotely is invisible to the lecturer and hence cannot influence their delivery.[149]

3. **Expressed commitment and injection of energy**: In discussions about the value of lectures, we have heard students say that it is like going to the cinema rather than watching the same film at home. By this they meant to draw attention to the collective and ritualistic aspects of attending a lecture, through which students come together and constitute themselves as 'students in this university on this course' and not as atomised, individual learners. What we see in the lecture audience may not be as intense as the 'collective effervescence' of a religious service or rock concert, but aspects of this kind of mutual encouragement and assessment of value are still present. The creation of a shared experience has value in learning and can also lead to discussion and the growth of mutual comprehension and shared understanding of what is important and trustworthy and what is not, in social interaction among students when the lecture is over.

---

[149] At least some lecturers find it hard to perform well in front of a 'dead' audience just as would entertainers of various kinds.

In summary, when looking for the benefits of the face-to-face lecture as com-pared to the recorded version of the 'same' thing, it is only by looking at the tacit and socialization elements that we can see what is missed by focussing solely on the explicit and this, again, helps to explain the 'stickiness' of even such a diluted form of the face to face as the 'formal' lecture – it is not quite so formal after all!

Conclusion to Chapter 7

What we have tried to show in this chapter is that even in groups and organizations where there seems to be every advantage to relying on remote communication, and no cost in terms of provision of the technology, face-to-face communica-tion does not go away. Face-to-face communication is, as we have put it, 'sticky'. We have tried to explain why it is so sticky drawing on the arguments of earlier chapters. We now return to 'the logic' of the case and its consequences.

CHAPTER 8

When Remote Communication
Is Not Trustworthy

The illusion of intimacy: remote communication looking local

In Chapter 6 we looked at the way what appears to be trust in remote persons is often trust in local reports of previous experience; this resolved some of the appearance of paradox found in our apparent readiness to trust distant strangers. But some of our trust in distant strangers is not so much paradoxical as sinister, because the way these strangers appear over the internet is in the form of a local community. This is 'the illusion of intimacy'. It can happen because the very characteristics of local communication that we have described in earlier chapters can be reproduced by social media and the like. Thus, while our overall argument is about the importance of local communication in science and the importance of the institution of science in good decision-making of all kinds, and therefore the importance of preserving the face to face, we are now going to embark on showing why people's trust in the face to face is misplaced where the intimacy is an illusion.

Much of the effect comes from the social organization of communication as well as the disguising of the communicator. It is the very power and central role in society of *face-to-face* communication that makes this illusion so insidious.

The societally dangerous illusion of intimacy arises in a number of ways which we will explore in detail in Chapter 9, using examples of disinformation campaigns.[150] To summarise, the illusion arises in organised disinformation

---

[150] Rebecca Lewis (2018) has investigated the spread of influence and popularity of 'alt-right' content producers on the video-sharing/social-networking platform YouTube. According to Lewis, alt-right YouTube influencers 'build trust with their audiences by stressing their relatability, their proximity,

How to cite this book chapter:
Collins, H., et al. 2022. *The Face-to-Face Principle: Science, Trust, Democracy and the Internet.* Pp. 141–150. Cardiff: Cardiff University Press. DOI: https://doi.org /10.18573/book7.i. Licence: CC-BY-NC-ND 4.0

campaigns through the use of strategies such as 'narrowcasting', which tailors the message to the individual in the way that normally happens when people are well known to the sender, creating an ersatz sense of homophily. It arises from 'inverse broadcasting', which constructs a seeming group of people similar to the target of the communication, the kind of group that is normally only found among friends and acquaintances or, in older times, fellow tribespersons, to reinforce the false message – homophily again being involved and reinforced by the sheer number of contacts that are typical of local groups rather than distant contacts; this enables weak links to simulate strong links, both in terms of moral strength and structural strength. As we will see, homophilous tendencies are easily faked still more directly by the construction of false personae, and media which readily transmit images and are ideal for this purpose.[151]

Social media can also give the impression of proximity by the sheer frequency with which messages are exchanged – a long way from letter-writing or even email – and much more like face-to-face communication among friends than advertising and propaganda coming from a distance.[152] This frequency of communication, when it is two-way rather than one-way, could build the sense of

---

their authenticity, their accountability and their similarity to those audiences... [and] by providing a specific social identity for themselves and their audiences.'

Audience members can contact content producers directly to critique or compliment their content whilst content producers distance themselves from the markers of authority – professionalism, distance, objectivity – adopted by traditional media outlets – they cast themselves 'as people *just like us*'. But '[t]hese narratives, while compelling, often obscure the skills, capital and networking opportunities these influencers gained from previous institutional affiliations. And many political influencers still maintain institutional affiliations to academic institutions, think tanks, and media outlets' (p. 17).

The techniques allow for the consolidation of a digital community around alt-right YouTube content, built on a shared social identity 'for those who feel like social underdogs for their rejection of progressive values, [whilst] provid[ing] a sense of countercultural rebellion for those same audiences' (p. 15).

[151] More evidence for the way the internet encourages a sense of intimacy is the readiness with which vulnerable people can be remotely 'groomed' or persuaded to send gifts to favoured personalities.

[152] Collins can report that he finds trust in distant relationships – such as with editors or publishers – tends to become fragile when emails are not answered in the regular way; face-to-face communication has a dependable and immediate rhythm, and this kind of rapidity and dependability is important if a distant relationship is to be maintained and to pass as something more intimate.

a community of the like-minded – just like the family or tribe. Add to this the use of emoticons and emojis which can be read as substitutes for friendly body language; in addition, there is the use of affirmatory signalling such as 'likes', sharing, commenting and other forms of approval.

The techniques of making strangers seem like friends have been known from long before the advent of social media. They are, for example, the stuff of confidence tricks, and then there are the handshakes and other physical touches and gestures which are commodified by politicians and public functionaries, creating a simulacrum of trustworthy familiarity. An iconic example is the 'politician's handshake', using two hands to enclose the other's hand, suggesting close intimacy. And of course, local groups of friends could act as echo chambers long before the invention of social media, and the echo chamber and filter bubble effect are reinforced when content appears in people's feeds precisely because it is 'liked' and shared by their actual friends and family – genuine local group links are intermingled with distant links simulating local links, building the overall illusion. A vast number of false friends can also be generated – the followers whose presence and whose influence can be engineered by techniques we will see in the Russian case to be discussed in Chapter 9. And some of those seemingly drawn into the affirming circle may be celebrities (who themselves sometimes *buy* extra fake followers).

As intimated, it is precisely the knowledge-forming power and trustworthiness of face-to-face communication that creates the conditions for its subversion in confidence tricks and power-plays, that also creates the conditions for the illusion of intimacy in social media and its internet bedfellows. Ironically again, Centola's experiments, intended to reveal features of networks in general, have shown us just how convincing these simulated groups of buddies can be.[153] If these remote media can mimic what happens to such positive effect in local forums, they will accrue the power of what happens in local forums but, since the connotations of trustworthiness are simulated, it is potentially damaging to the fabric of society. And the same communication techniques render the fabric of society liable to damage and distortion by causing chaos and liquefying reality – making it ready to be shaped by organised communicators. Once upon a time massive social change was very hard to bring about because of the obduracy of taken-for-granted reality – war and social revolution were the only

---

[153] The point will be explained in more detail below. Eliasoph (1998) argues that Americans avoid deep political arguments in public and reserve that kind of principled argument about their political intentions to the 'backstage', among trusted acquaintances. It could be that the switch to remote communication could, then, switch off deep political debate as these trusted forums no longer exist. We would argue that this is not necessarily the case given the illusion of intimacy but that if the debates did continue over the internet they would be open to untrustworthy influences.

liquefiers; with the growing power and ubiquity of the internet, liquification of knowledge is becoming the default, and the problem is how to stop it.

Part of the explanation of how this works is the ease with which anonymity of sources can be achieved and disguised; this allows an organised source of influence to be passed off as something else, such as a neutral news agency. For instance, *The Guardian* of 2 August 2019 (pps. 1, 12, 13) reveals that the firm CTF Partners (closely associated with Boris Johnson's campaigns) which represents various interests such as coal companies, Saudi Arabia and African countries, has promulgated media campaigns under such assumed 'news' identities as 'Middle East Diplomat', 'Iran in Focus', 'Inside Mauritania' and 'Free and Fair Election Zimbabwe'. But when, on top of this, the face-to-face interaction that is responsible for our acquisition of a sense of the real through both primary and secondary socialization can be inexpensively mimicked *en masse*, we have created a malleable social fluid from what was once our foundational reality.

Networks revisited

An additional sense of how this works can be gained by going back to the analysis of networks that was introduced in Chapter 6. Centola argues that the wide bridges needed for the spread of *ideas that require commitment* will be associated with local groups. And this also fits with Granovetter's later (1985) claim that successful economic networks, far from being represented by the purely self-interested and perfect information models of mainstream economics, depend heavily on personal trust relationships and networks both within and between firms. And it fits with our own case studies both of economic relationships and frontier science.[154] Centola also argues, supported with reports of earlier empirical studies, that his 'complex contagions' spread best by what we have called 'social diffusion', in the way we have suggested accounts for the societal uniformity of natural languages: close-knit, families are the sites of linguistic socialization, but they influence and are influenced by other families, and so on. The complex contagion – it could equally be a demanding way of thinking or acting – develops best in local groups then diffuses across the boundaries to other local groups, and so on. This is a much better way of spreading complex new ideas, requiring moral trust, as we would say, than distant communication over remote weak ties.[155] But, in the terms used by Granovetter and Centola, weak links, which are remote links, can be made to look like local links – like

---

[154] Granovetter (1985) and see the study of 'Huge Stores' in Chapter 7.

[155] Centola also discusses, on pps. 38–40, the psychological mechanisms that encourage the adoption of ideas when the individual is surrounded by many other adopters. This is relevant to understanding the impact of social media though in this book we look at it primarily through the perspective of trust.

emotionally strong links. Then, by artificially multiplying the number of weak, remote, links, they can be made to look like a wide bridge. That means that weak links will be accepted as transmitting the kind of knowledge that needs moral trust, perhaps difficult and dangerous trust, not just information. In Centola's terms, this can give rise to a complex contagion.

That this is the case is pretty obvious even without the theory or the kind of detailed analysis of the data we have seen above. Many of the techniques used to make social messages persuasive are not new; they have long been used in advertising and propaganda.[156] But, on the whole, foreign propaganda does not have much impact. For example, the immensely powerful and skilful propaganda of Nazi Germany had little impact on Britain because it so obviously emanated from a foreign country; the leader's rants, which might have been persuasive to some, were in a foreign language. Anonymity and the creation of a false identity and personality require enormous artifice when interaction is local, but when multiplication of sources is added to disguise drawing on the anonymity which is characteristic of the internet, the effect can be potent.[157]

Broadcasting, narrowcasting and inverse broadcasting

Two important ways of organizing social media campaigns are, as already mentioned, narrowcasting and inverse broadcasting. Narrowcasting refers to adver-

---

[156] And see Appendix 1. Of course, since at least the 1970s, in, for instance, the work of Stuart Hall and other Cultural Studies scholars, media content has been argued to be open to multiple interpretations or 'polysemic'. Different audiences can and do interpret media content differently, based on demographic and attitudinal factors (e.g. Hall et al. 2003; Kitzinger 2004; McQuail 2005), making the specific impact of media content hard to predict or measure. This remains an ongoing debate amongst media theorists, but the large quantities of money that continue to be spent on mass-media advertising suggest that the belief that media content does have somewhat predictable effects on audiences' understandings and perceptions is widely held, at least by those in positions of relative social influence.

[157] Thus, to take a current example, if 'Texas Lone Star' (Figure 9.1) announced himself as a Russian living in Russia, the impact would not be boosted by apparent homophily but eroded by nationalistic difference – more especially if the language was Russian. Erving Goffman's descriptions of 'The presentation of self in everyday life', going back to 1959, describe a world in which ordinary human life is theatre and humans are continually constructing new personae for themselves for others have to interact with. Manipulation and deception were key themes in Goffman's work, and his insights into deception in face-to-face encounters can also help explain how prepared we are to pick up false personae from the internet.

tising and political campaigns which are designed to be effective in respect of specific sectors or members of a population whose characteristics have been harvested through their general internet activity or social media profiles.[158] Being tailored to specific kinds of individuals, the manufactured links can appear to be strong rather than weak in the emotional – homophily – sense. Inverse broadcasting is the attempt to influence the views of single persons by targeting them with social media messages from what appear to be many independent individuals – strong in the wide bridge sense – but whose seemingly spontaneous approach is managed by a single source. Actually, in inverse broadcasting the process is repeated so that many single persons are targeted, so it is really broadcasting presented as its opposite: each target experiences the approach as though they were a unique 'friend' being contacted by many other 'friends'. The flow in inverse broadcasting appears, therefore, to be many-to-one and this can be influential. We will see examples of both narrowcasting and inverse broadcasting techniques in the Russian disinformation campaigns discussed in Chapter 9. Inverse broadcasting is the opposite of what happens in normal broadcasting and of what happens in Blockchain technology: in both cases what starts with individuals – either talk or a transaction – is transmitted to, or guaranteed by, its witnessing by many individuals – the flow being from one-to-many.

Figure 8.1 is meant to capture these features of internet communication in network diagrams, some of which are specific to internet communication. In Figure 8.1, all solid links are taken to be unidirectional downward-pointing communication channels while dotted lines, when we get to them in panel E, are bidirectional.

Panel A is broadcasting of any kind but is there to contrast with narrowcasting, which is largely confined to the internet (though one may imagine political leafletting directed to regions of towns known to be inhabited by voters of a certain persuasion, being a kind of 'clunky' narrowcasting, as with the different kinds of adverts found in newspapers and magazines with a politically identifiable readership). The difference between broadcasting and narrowcasting can be seen in panel B, with different messages being transmitted to those of different persuasions (represented by different shadings). In panel C, the set of individuals doing the targeting is centrally organised, the organiser being represented by the large black disk so that many individuals offer a similar message to the recipient – an artificial wide bridge. In panel D this is repeated over and over again – each set of small black disks being the same people. The number of people targeted can be multiplied indefinitely as indicated by the dashed line. There is no reason why inverse broadcasting should not be combined with

---

[158] For 'narrowcasting' see https://dl.acm.org/citation.cfm?id=2492570. Pomerantsev (2019b, p. 10) quotes Thomas Borwick, the director of the UK's Brexit favouring Vote Leave Campaign as saying that 70 or 80 types of targeted message are needed for a country of 20 million voters.

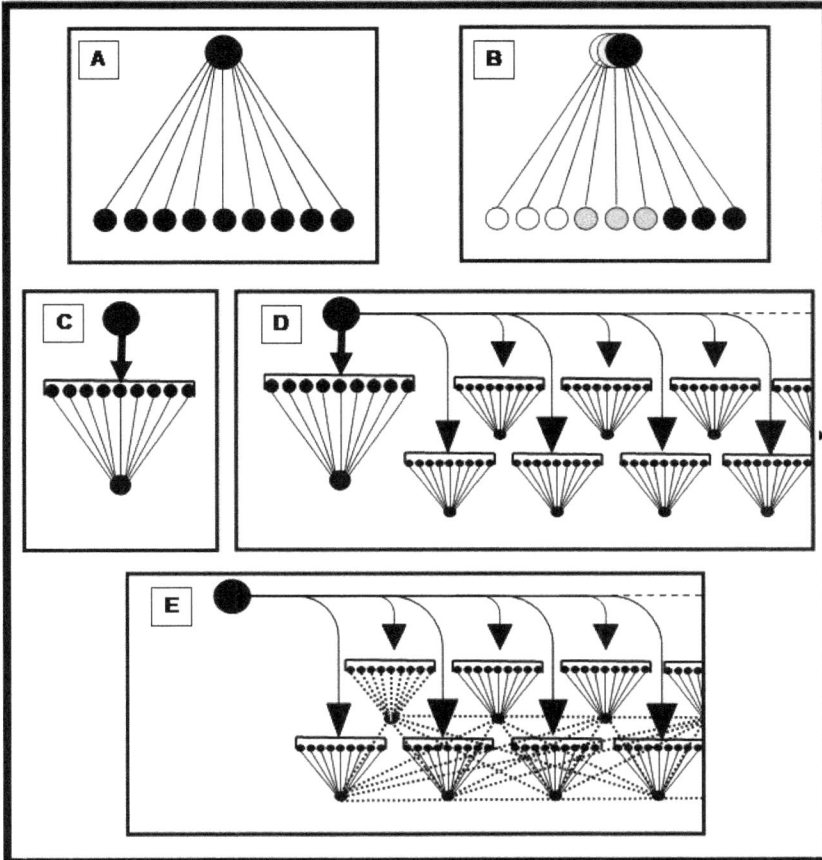

Figure 8.1: Broadcasting (A); narrowcasting (B); inverse broadcasting (C + D); community building (E).

narrowcasting to target distinct groups, giving distant weak ties the appeal of both homophily and a wide bridge.

Because of the way social media is used, individual users can find themselves talking to each other and building what feels like a community, as illustrated by the bidirectional dotted links in panel E of Figure 8.1. That community, in some cases, will have been constructed by the single interest represented by the large black disk in the figure. Of course, a necessity in every case is concealment of the true source of the messages – to repeat, this is how the extremely efficient, long-distance, weak ties can be disguised as a local wide bridge.

The sense of community will be strengthened if there is back-and-forth communication between the individuals at the bottom and the many, apparently independent communicators represented by the nine smaller black disks at the top of each triangle. As far as we know this does happen occasionally if not regularly, but one can see the possibilities for the future. If the unidirectional links are changed to bidirectional, as illustrated for the individual labelled '1' in panel E, 1's community increases from 'eight' others (we use inverted commas to indicate that the numbers are nominal), to 17 others but with '9' of those others actually being remote and remotely controlled 'disinformers' – for example, they might well be Russians living in Russia. Note that in this scenario, the same 'nine' persons are linked into each of the other 'eight' so they know what the other 'eight' are writing and can respond to any one of the 'eight' as though there are a member of the whole community. Imagine if, during the 1930s, what one took to be one's community, contained a preponderance of Nazis resident in Germany! Or imagine that in the 2010s in Britain, many of one's supposed community were 'fronts' for a centrally organised campaign by persons with a self-interested determination to bring about a Brexit irrespective of the public interest!

Centola's experiments on social diffusion and their unintended meaning

Centola's study of social diffusion is remarkable in that he actually carried out some ingenious experiments on communities that he created in order to check the theories he had developed in the first part of his 2018 book. He wanted to find out if networks that had local clusters and no weak ties were better at encouraging changed behaviour than those where the contacts were randomised across the whole network. Centola set up a series of experimental health advice networks using the internet, with the linkages in various networks being differently arranged. The individuals taking part in these networks were anonymised and represented by avatars. The results of the experiments are clear and support the theory of wide bridges. Where joining the network involved some commitment of time and effort, locally clustered networks drew in more participants, and drew participants in faster than randomised networks, and membership spread through the networks via what we and Centola call social diffusion.

Centola is well aware that this experiment did not exactly reflect what happens in non-internet-based communications (see pp. 82–83 of his book) but he presents these experiments as an investigation of social networks in general, claiming, not unreasonably, that the effects he documents would be still more marked in the real world where the other group-reinforcing social psychology effects were active. The only reason his experiments used anonymised networks based on interactions over the internet was because this was the only way to do them. It is relatively cheap and easy to set up such networks and easy to monitor how they develop; it would have been logistically impossible to set up various kinds of face-to-face networks in the real world and completely impossible to monitor the development of real-world interactions since one would have to monitor participants' behaviour 24 hours per day. These novel and brilliant experiments supported the theories he was putting forward.

In the light of this book, however, the striking thing is that Centola does not make the relationship between face-to-face contacts and remote contacts the salient point. His experiments dealt with remote contacts mediated by the internet but, to repeat, he treats them as representing social networks in general. In the light of our arguments, they do represent social networks in general, but not because face-to-face and internet communication are identical, but because certain kinds of internet communication *mimic* face-to-face communication. Centola's experiments can be seen as an investigation of the illusion of intimacy!

Centola selected health advice networks to experiment with because there are a number of precedents and he is able to report how 'well they work' – or as we would say, how well they *mimic* face-to-face communication – in the following terms:

> I was struck by the sincerity and commitment that participants on health sites exhibited in their interactions with strangers. A staggering wealth of sensitive medical data ... is exchanged in settings – including medication details, health diaries, MRI and CAT scans, medical reports and physician referrals. Most of this information is uploaded, shared, followed and commented on by people who have never met face to face. (p. 66–67)

He says 'the social interactions on these websites were surprisingly natural'(p. 68). He says '[the experiment] was designed to be a natural setting for participants to have interactions with strangers who might influence their behaviour'(p. 69).

Reinforcing this sense of intimacy among the participants, he refers to the interacting immediate internet neighbours in his experiments as 'health buddies'. Yet we know from our analysis of what happens in scientific core sets, the Checkov case study, and so forth, that interactions that require the level of trust needed to develop new physics knowledge have to be face to face or based on a history of face-to-face interaction. To repeat, Centola's experiments, as we would see it, are illustrations of how the *illusion* of intimacy works: they make participants who are interacting with anonymised persons over the internet represented by avatars feel like 'buddies' belonging to local groups.

The strength of this feeling is indicated, because, as Centola says: 'these interactions [the anonymised interactions in his experimental trials] can influence patients' decisions to take medications, join medical trials, and change physicians' (p. 67), and these are 'difficult and dangerous' matters.

That this is true shows the power of the internet at mimicking local groups. Remember, as Centola says, 'The strategy was to bring people into the study in a way that would get them excited about their immediate social contacts but not give them any information about the scale and structure of the larger social network' (p. 69). This might well have been referring to the point we have been making about the invisibility of the difference between remote participants and local participants in disinformation campaigns. Even though Centola's experiments involved well-meaning and trusting people and were directed at doing them good, they could equally apply to the diffusion of anti-vaccination campaigns, or state-sponsored disinformation campaigns conducted over the internet. What Centola's experiments show is just how careful we must be with this kind of internet campaign. His clustered networks, the most effective ones, were also the most effective at creating an illusion of intimacy.[159]

---

[159] This is not to say that these networks were not working beneficially, just to say that it was illusory to imagine that others in the network necessarily had the same commitments as face-to-face buddies.

CHAPTER 9

Disinformation and Misinformation

We now present some examples of how the illusion of intimacy works and how hard it is to combat, starting with the example of use of state-sponsored disinformation in Russia.[160]

An example of the malign use of the internet: Russian disinformation techniques

The data

The Russian 'Internet Research Agency' (IRA), based in St Petersburg, first achieved public notoriety amidst the swirl of allegations and accusation that the Kremlin had sought to interfere with and influence the outcome of the 2016 US Presidential election. Several high-profile and well-publicised reports have documented aspects of their activity.[161] New empirical data shows that the IRA's assets were first mobilised 'at scale' for a domestic audience around 2012, to help shore up the popularity of President Putin amongst Russian citizens. They were subsequently directed towards helping counter the geopolitical 'fallout' of the downing of the MH17 airliner in Ukraine in 2014, and the annexation of Crimea by Russian forces that same year.[162] The release of data in 2018 by

---

[160] The data on Russian disinformation campaigns was gathered by Martin Innes and his colleagues. The analysis reported here draws in particular upon work by Andrew Dawson, Kate Daunt, Helen Innes and Diyana Dobreva and was, in part, supported by funding from the Centre for Research Evidence on Security Threats and UK Government.

[161] For example, Mueller (2019) and Digital Culture, Media and Sport, House of Commons Parliamentary Select Committee (2019).

[162] Dawson and Innes (2019).

How to cite this book chapter:
Collins, H., et al. 2022. *The Face-to-Face Principle: Science, Trust, Democracy and the Internet.* Pp. 151–162. Cardiff: Cardiff University Press. DOI: https://doi.org /10.18573/book7.j. Licence: CC-BY-NC-ND 4.0

several social media companies make it possible to reconstruct aspects of the IRA's operating procedures and methodologies.

It appears that it was public protests in Russia between December 2011 and July 2013 against what were perceived as flawed elections results – sometimes known as the 'Snow Revolution' – that was the catalyst for the IRA's initial attractiveness to the Russian government.[163] In the middle of 2013, when the protests were waning, the IRA started to create large numbers of English language accounts – over 400 in August alone. This is suggestive of there being strategic shift in interest within the Kremlin towards targeting the Agency's assets towards foreign issues and interests.

However, at the end of 2013 protests broke out in Ukraine. Events escalated quickly with: (1) the appearance of the so-called 'little green men' (soldiers devoid of any insignia) in Crimea; (2) its annexation in an illegal referendum; (3) the shooting down of flight MH17; and (4) the war in the Donbass, all these occurring within six months. Over 800 new Russian language social media accounts that have been linked by Twitter to the IRA were also set up in those six months; as the salience of events in the Ukraine faded, attention was once more turned to external influence. In 2016, 185 new accounts were created: 50 Russian; 1 Spanish; 1 French; 62 German; and 71 English. Whilst the US was the primary target of these accounts, as can be seen, German public opinion was also a significant focus. The tweets were searched for names and leading politicians running in German elections in 2016: Henkel, Müller, Meuthen, Kretschmann, Lederer and Sellering were found.

It seems that at least some of the German activity was intended simply to sow discord as conflicting narratives were promulgated in these tweets, for example, stances on both sides of the high-profile debates around Syrian refugees and Chancellor Merkel's policy relating to it. On the other hand, the IRA trolls were united around promoting opposition to President Erdogan and Turkey in general; jokes were made about Erdogan planning the refugee crises; they also promulgated fear about Turkey joining the EU and the consequent movement of people.

Similar narratives were detected when the dataset was searched for the names of Austrian leaders: many tweets by IRA-linked accounts highlighted crimes committed by immigrants and blamed the electorate for choosing 'more murders, rapes, etc' as a result of electing Alexander van Der Bellen. The narratives for Italy had more focus on its relationship with Ukraine/Crimea and its support for blocking sanctions against Russia. There was evidence of support for the far right '5 Star Movement' and its leader, Matteo Salvini. Although national elections were not held in Italy that year, there was IRA support for the referendum vote, using the hashtag '#bastaunSi!'.[164] The IRA-linked accounts

---

[163] Wikipedia (2021b).
[164] Meaning 'Just a yes!'

Figure 9.1: Texas Lone Star – a Russian spoofed identity and account.

also repeatedly messaged that Italians should leave the EU, scrap the Euro and quit NATO, in order to recover their sovereignty.

The method

The tweets had a chance of being impactful in the US because operators in St Petersburg were able to mimic and project the social identities of certain kinds of American users of social media. They appropriated symbols of identity and constructed fake personas. For example, Figure 9.1 shows the avatar and biographical self-descriptions of an (in)famous of IRA account (original in colour).

Other IRA accounts generated different personae, such as 'blacktivist' accounts, with equally clear markers of identity.[165] What is being exploited here is the concept of 'homophily': people are likely to attend to messages from others who seemingly 'look and think' as they do and this also gives rise to the underpinnings of 'echo chambers' and, to a certain extent, 'filter bubbles'.[166]

---

[165] There are echoes here of Erving Goffman's (1961) concept of 'identity kits'.

[166] Individuals respond to and are persuaded by sources in social networks that are perceived as similar to themselves, likeable and credible (McPherson et al. 2001). In the context of social networking sites, Chu and Kim (2011) report evidence suggesting users believe that homophilous sources are more

But 'identity spoofing' is only part of the game. Some of the IRA accounts were investigated in detail and showed extended messaging histories. Operators consistently tweeted about topics coherent with the spoofed account identity, slowly and patiently increasing the number of followers over a period of several years, even though no political content was being transmitted in this primary process of building a homophily-based network: socially exploitable relationships were being accumulated so that at some future moment a political message relating to previously unanticipated high-profile events could draw instantly on the already charged-up 'social battery'. For example, 47 IRA Twitter accounts were found to have sent messages to their followers in the aftermath of the series of terror attacks that took place in the UK during the first half of 2017. Eight of these did so repeatedly and achieved a large number of audience engagements with their messages.

Patiently building social potential was not the only method. When possible, especially as part of collective reactions to high-profile events, the IRA operators would amplify and get behind messages originally sent by other social media users so as to avoid having to create their main message *de novo*; this presumably, was both quicker, required less effort and reduced the chances that the form of the fake message might contain a mistake or error.

Some operators preferred to use 'shortcuts' that 'hacked' Twitter's algorithms. Some *bought* followers after the manner of certain celebrities' social media accounts.[167] Increasing the ratio of followers to followed appears to have both social and algorithmic effects. The more followers the more the operator's credibility appears to be enhanced by the affirmation of the similar members of the 'tribe', what Innes et al. label 'social proofing'.[168] But Twitter's ranking algorithm also responds to such metrics, affording an opportunity for a skilled operator who can manipulate such data to inflate the visibility of their messaging.[169] In typical successful Twitter accounts, there is a relatively steady increase in

---

trustworthy and reliable than other ones. Humans want social acceptance by group members and alter their attitudes, opinions and beliefs to adhere to group norms and gain/maintain social acceptance and approval from others (Nahai 2017). Analysing this drive in the context of social influence, Cialdini and Goldstein (2004) show that group pressures can easily alter an individual's moderately opposing attitude towards the group norm, to the position of adopting the wider group consensus (i.e. the group can successfully alter the group member's personal attitude towards an issue). This was, perhaps, most famously first demonstrated by the perception of Asch's (1951) perception of line length experiments discussed earlier (See Chapter 3, Small face-to-face groups and science).

[167] Marwick (2013).
[168] Innes et al. (2021); see also Cialdini (2009).
[169] Margetts et al. (2016) are very incisive on these social and technical affordances.

followers over time. In accounts where followers are purchased, exemplified by what Twitter Inc has labelled an 'Internet Research Agency linked account', there is a vertical step change where the account operator has purchased 10,000 followers. There are multiple sites online offering this service, with the price reflecting various 'promissory' notes about the 'quality' of the following accounts that will be provided.

A second short-cut technique can be called 'fishing for followers' and involves making an account more visible to other users. It is based on Twitter's proprietary 'ranking' algorithm and works as follows. First the operator 'follows' a large number of accounts and indicates to their owners that they should 'follow back' in return. Then a few days later, the spoof account 'unfollows' all these, hoping the unfollowing move will not be noticed by the majority of users. The result is inflation of the ratio of those who follow the account to those who the account follows. This creates the impression, as signified by Twitter's ranking algorithm, that it is an important 'influencer account'.

A third technique, the purposes of which are not immediately obvious but which show up unambiguously in the behavioural patterns of known IRA accounts, is 'narrative switching'. This involves an account with a particular identity and a messaging content that coheres with it, suddenly and dramatically switching position. For example, a German language account that started out quite explicitly anti-Angela Merkel in its stance, then falls dormant for several months, before re-surfacing, but now vociferously supporting Merkel's policies and government. This could be simply intended to cause disruption by de-stabilizing fixed views and giving rise to conflict in the community – further liquefying social reality and readying it for manipulation; it could be a way to animate an account and encourage users to engage with it, something of great value to those seeking advertising revenue and likely of great value to those who want engagement for political purposes; or it could be a convenient way to address a ready-made audience for whatever set of ideas it is now thought beneficial to promulgate. Of course, the messages will not be as immediately appealing as to an audience with similar views, but they can still have various other kinds of impact including keeping those with every view engaged.

Since the IRA accounts often do not originate material but amplify messages originally authored by non-IRA accounts where these fit with the Agency's focus and strategy, another 'low cost – low effort' method is available: this is the profligate use of 'bots' to amplify key messages directed at audience segments. Many IRA accounts were bots – sections of autonomous computer code. For example, '@News_Executive' was an IRA bot that repeatedly pushed news stories in the direction of its followers. Though the majority of the known IRA bots seemed relatively ineffective, they are so cheap and easy to produce and deploy that their collective effect, in terms of 'polluting' the media ecosystem, might well be significant.

Analysing the behaviour of 2,800 IRA Twitter accounts shows that they have similar 'pulses' of messaging activity – in other words, these accounts seemed to act in a coordinated way, not as the work of isolated individuals. This suggests that multiple accounts were being directed by 'managers'.[170] More than 70% of the IRA accounts had a very similar activity profile to at least one other account. Applying the analytic procedure to the whole dataset shows the existence of clusters of accounts. This corresponds to the reports of journalists who interviewed former IRA employees and found it to be organised in a series of departments, specializing in specific countries and platforms. The biggest cluster was focussed on Russian speakers with smaller, 'satellite' clusters operating in different languages, including French, Spanish, German and Italian, whilst also assuming particular ideological positions (e.g. far-right/far-left). Such an English language cluster can probably be identified with 'the American Department' responsible for activity related to the 2016 US Presidential election.

Instagram, images, and interactions

Instagram has been studied less than Twitter and Facebook, but a US Senate investigation into the IRA's activities in relation to the 2016 US presidential election suggests Instagram was an important part of the Kremlin-backed influence and interference campaign. It suggests it may have had a greater impact than the more well-known Twitter activity.

Instagram's popularity has exploded since Facebook purchased it in 2012; while younger users are leaving Facebook, Instagram is in vogue. Instagram is used by 71% of American 18–24 year olds compared to Twitter's 45%.[171] Instagram is the second most popular social media platform for teenagers and is seeking to usurp Snapchat by copying its most popular features.[172]

On 5 November 2018, a day before the US midterm elections, Facebook released a statement announcing they had removed 30 Facebook accounts and 85 Instagram accounts (later revised to 99 accounts) for engaging in 'coordinated inauthentic behaviour'. This followed upon a tip from US law enforcement agencies suggesting that these accounts were connected to foreign governments.

Many regular users of social media adopt multiple identities to run their different accounts so the term 'inauthentic' is being used here to connote something more. According to Facebook, around 1,250,000 people followed at least one of these deleted Instagram accounts, giving an average of 12,000 followers per account. Because Facebook removed most of these accounts before the account names were made public, it was not possible to archive the data. However, thanks to the large numbers of third-party websites offering 'Insta metrics', the researchers were able to archive an average of the last 12 posts

---

[170]   Possibly using a system like Tweetdeck.
[171]   http://www.pewinternet.org/2018/03/01/social-media-use-in-2018/.
[172]   https://www.recode.net/2018/10/9/17938356/facebook-instagram-future -revenue-growth-kevin-systrom.

made by each of the accounts and show that they were focussed towards a left-leaning political audience, specifically those engaged with identity politics. A breakdown of the left-leaning accounts shows that 'blacktivism' was the most popular identity category (n=15), followed by LGBT and Feminism (n=5 each).

Black cultural icons were heavily represented in the Instagram celebrity category, with Rihanna and Kendrick Lamar being the most popular. This focus on black issues may be the IRA simply responding to demographics: 43% of black respondents to a Pew Survey stated they used Instagram vs. 32% of white respondents. Alternatively, it could be that they are rehashing tactics from the Cold War, trying to counter US hard power by distracting them with domestic issues around inter-ethnic tensions. The key point though, is that this political profile is markedly different from what has been reported in respect of the IRA's 2016 Twitter and Facebook operations, where most effort has been directed to mimicking and manipulating radical, far-right, thought communities.

An account targeting Ukrainian issues exemplifies the way IRA operators worked with Instagram. Suspicions about it were aroused because it propagated a meme regarding Natalia Poklonskaya: Poklonskaya was the Senior Prosecutor of the Prosecutor General of Ukraine under Victor Yanukovych, and became famous when she was appointed the Prosecutor of (the autonomous/occupied) Crimea. A video of a press conference she gave to mainstream media became a viral hit because of her attractiveness, with jokes like 'Steals Crimea, and your heart'.

The account, *Kasarov_eli* (now removed) was presenting as an anime fan account and they had posted several drawings of Natalia in an anime style, as illustrated below in Figure 9.2.[173]

Periodically placed amongst the messages posted were repeated tagging of pro-Russian words, such as #VladimirPutin. These were sometimes buried within lists of as many as 30 hashtags. As noted in the previous Twitter analysis, this technique appears designed to 'game' the platform algorithms in such a way as to maximise the reach of the account's messages.

Figure 9.2: Pictures believed to have been posted by the Kasarov eli account based on google cache.

---

[173] Anime is a particular style of Japanese animated cartoon drawing that has spawned a very large and devoted online subculture.

The account also reposted statements from the official Instagram account for the Presidency of the Syrian Arab Republic and Ramzan Kadyrov (the Head of the Chechen Republic and currently sanctioned under the Magnitsky Act for involvement in repression, torture and murder) – not something typical for your average anime fan! Especially striking was how the account evidenced a roughly 80:20 ratio in terms of pro-Russian content. That is, around 80% of its posts were anime and/or largely unremarkable, but 20% focussed upon topics and propaganda themes of interest to the Russian state. As the history of this account was 'unpacked', it transpired that it had 'cycled' through a number of name changes and aliases including: *kasarov_red*, *spooky_kasarov_eli* and more intriguingly *god_hates_kebab* (kebab being internet slang for Muslims in some communities), *united_russia_republics* and *stop_russophobes*.

As well as drawing on the tendencies associated with homophily, social media, such as Instagram, with their ability to convey pictures as well as words, use all the traditional techniques of advertising and propaganda, triggering emotion and motion to capture user attention. These resources meant that IRA workers based in St Petersburg were able to get Instagram users around the world to engage with the messages they were sending, reinforcing the message with apparent high levels of group endorsement as described above: as various studies have demonstrated, users translate high levels of endorsement as a proxy for credibility.[174] Accounts also benefit by association with a celebrity.[175] Multiple IRA-created accounts were oriented around a range of different celebrity figures, some illustrative examples of which are reproduced in Figure 9.3.

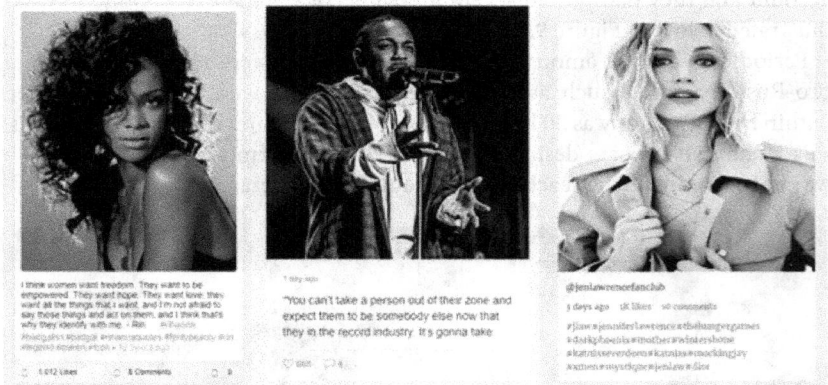

Figure 9.3: The use of celebrities to attract and influence social media users.[176]

---

[174] Jaakonmäki et al. (2017); Veirman et al. (2017); Coulter and Roggeveen (2012).
[175] Muñoz and Towner (2017); Fu et al. (2017).
[176] The account names and number of followers at the time of the research are listed below:
https://www.instagram.com/Riri_one_love (80,000 followers); https://www

Figure 9.4: The use of visual cues in black activism posts.[177]

Visual cues are also used in the case of black activism accounts / posts in Figure 9.4.

Repeated exposure to a message has also been shown to be a strong determinant of users' awareness of, engagement with and agreement with that message. That is, when individuals see or hear an argument repeatedly, they are more likely to remember its content and rate it as 'true' compared with unfamiliar statements.[178] Applying this logic to Instagram, users who are recurrently exposed to the same argument within a single account, or across multiple accounts, are likely to remember those arguments and find them credible.[179] None of this is surprising given the contribution to socialization of the bath of words in which we float.

Initial conclusion on social media

The above analyses show that social media operators thousands of miles away in St Petersburg, and with supposedly only a relatively 'surface' understanding of the cultural norms and conventions associated with the social orders and realities in which they were actively intervening, can effectively deceive

---

.instagram.com/Kendrick_dna_(86,000 followers); https://www.instagram .com/Jenlawrencefanclub (34,300 followers).

[177] The account names at the time of the research are listed below: https://www.instagram.com/Black.dollar; https://www.instagram.com/Black .voices; https://www.instagram.com/blknation.

[178] Nahai (2017); Hasher et al. (1977).

[179] For an important update to this account of social media and disinformation, see Appendix 2.

and disinform. *Social media are able to blend intimacy with distance* generating the illusion of intimacy from afar.

Disinformation and science

There is evidence that Russian military intelligence has tried to hack the computer systems of British universities working on a coronavirus vaccine.[180] The interest in science goes back to the 'Cold War' and possibly even before, with nuclear science a particular focus; this is part of the very extensive programme of Russian 'active measures'.[181] In 2018, Russian intelligence agents were discovered in the Netherlands outside of the Organisation for the Prevention of Chemical Weapons building in the Hague with, seemingly, a hacking operation in mind, in possession of advanced surveillance and computer equipment. It was thought they were planning to undertake a 'closed access' hack operation using the local wifi network.[182] But sowing disinformation about HIV-AIDS had already proved so successful that it is thought that it still influences public understanding of the causes and consequences. Operation Infektion was a long-term effort by the Russian KGB to propagate the idea that HIV-AIDS was deliberately manufactured and spread by the US government. In an article published in the Central Intelligence Agency's 'house' journal in 2009, Boghardt explains how the Russian effort started with an anonymous letter in 1983 published in an obscure newspaper in India called *The Patriot*. The letter claimed that an American scientist and anthropologist had attributed the AIDS virus to a Pentagon programme experimenting with new biological weapons. As Boghardt describes, the letter lay dormant for three years until, in 1985, *Literaturnaya Gazeta*, one of the KGB's prime conduits in the Soviet press for propaganda and disinformation, published an article by Valentin Zapevalov, entitled 'Panic in the West or What Is Hiding behind the Sensation Surrounding AIDS', which drew on the letter. A retired East German bio-physicist Professor Jakob Segal, who, according to Boghardt, was known to the KGB and East Germany's Ministry for State Security (Stasi) took the story and mainstreamed it for several years, such that it became accepted as 'true' in some thought communities. It was also given a 'racialised' interpretation as intended to disproportionately target black communities, thereby drawing on and enhancing an existing social fissure. Russian intelligence services have an established playbook which often begins with faked reports in remote media outlets which might be picked up innocently by international media and republished as authentic or 'rediscovered' later.

---

[180] https://www.thetimes.co.uk/article/russian-actors-tried-to-influence -2019-general-election-says-dominic-raab-z57j6s825 (accessed 03/09/20).

[181] See Rid (2020).

[182] https://www.bbc.co.uk/news/world-europe-45747472 (accessed 03/09/20).

Unsurprisingly, the same techniques are being used at the time of writing in respect of the coronavirus (COVID-19) pandemic. In the main body of the book, however, we will use the MMR revolt as our principal example and the hard case against which we need to discuss a solution. The up-to-date COVID-19 disinformation material is set out in Appendix 2.

MMR disinformation

Innes's research team were able to provide original analysis of a disinformation campaign relating to the MMR vaccine, a case which we will discuss at length. This, again, originated with the IRA in St Petersburg. The large IRA Twitter dataset was searched for terms relating to anti-vaccine movements.[183] Searching the tweet text for the characters 'vaccin' found 2,109 tweets between December 2014 and December 2017. Only 467 (22%) of these tweets were retweets of other users' content, suggesting that the IRA were interested in creating their own tweets to drive this conversation. But the vast majority of these tweets are simply 'news' headlines about vaccinations in the US. These news headlines were tweeted by the IRA's fictitious 'local' news organizations in the US to provide authenticity to their accounts. The top 12 IRA accounts tweeting about vaccines were these fictitious news organizations purporting to be from places such as San Jose and Kansas. As such the majority view promoted by these tweets closely mirrors mainstream media (MSM), with many tweets noting the signing into law of California vaccine bill SB 277, without editorializing the issue. This bill removed 'personal belief' exemptions from vaccination requirements for schools and day care centres; this meant that in the future, parents who wanted their children to remain unvaccinated would have to home-school them or move to a different state.

Some accounts seemed to offer conflicting views on vaccination. Thus, one account described itself as a 'Conservative domestic goddess devoted to God, Family, Country, praying daily for #America! #tcot #ccot #RepealObamacare.' It tweeted twice in support of vaccinations saying 'If we don't have regular chek ups and get #vaccines-what's the point of doctors' work? #VaccinateUS'. Two minutes later, however, it tweeted that 'hospitalizations, irreversible brain damage, and hundreds of deaths that is what #vaccinations cause #VaccinateUS'. Interestingly this account went by the name of 'Jeanne Mccarthy' which is very similar to 'Jenny McCarthy', who is a US celebrity with an autistic child who has been described as the public face of the anti-vaxxer movement in the US.

Fifteen tweets in the large IRA dataset specifically mentioned the MMR vaccine. The IRA accounts that made these tweets had a combined following of over 162,000 accounts. Nine of these tweets were retweets. Of these retweets,

---

[183] This report on the IRA was, once more, compiled by Martin Innes and his team.

seven were promoting an anti-vaccine viewpoint, indicating in some way that the MMR vaccine should not be used, for example *'RT @LotusOak: Japan Banned MMR Vaccine From Their Schedule https://t.co/JftK35dIjE Infant Mortality in Japan 3x less than in US'*. These were primarily retweets from websites such as ActivistPost and WorldTruth. The other two retweets promoted a pro-vaccine viewpoint, for example *'RT @NYC Everyday: Another study finds no link between MMR vaccine and autism #health'*.

The remaining six tweets that mentioned MMR were authored by IRA operatives. Of these, five presented a pro-vaccine viewpoint, with all of the tweets mentioning a new study that found no link between MMR and autism. All five of these tweets came from accounts masquerading as 'local' news outlets from American cities such as New York and Boston. The remaining tweet was anti-vaccine, and implied that the MSM were 'fear mongering' to get the public to take the MMR shot.

This data gives some idea of the extent of Russian-instigated social media influence in anti-MMR campaigns.[184] This seems the right place to address one of the strangest features of the IRA's interventions in this domain: they are, very roughly, even-handed between pro- and anti-MMR. Why should this be? It seems to be not so much that the IRA want to encourage epidemics of measles but that they want to grow uncertainty and conflict. To intervene on both sides of the debate is to grow the debate and amplify the uncertainty: instead of vaccination being based on solid knowledge based in medical science and epidemiological assessments, it becomes a matter of opinion. There is, then, no need to change the substance of what people think by addressing it directly if one wants to control it. It may be sufficient simply to seed doubt and uncertainty because, once knowledge has become 'liquefied', the conditions have been created for someone other than its expert creators to control it. Where expertise in general can no longer be relied on, but instead becomes seen as no more secure than mere opinion or ideological preference, then the institutions of pluralist democracy which depend on expertise can no longer be seen as reliable, suiting Russian as well as certain Western leaders very well.[185] This is a topic to which we will return. That said, the anti-vax websites that are not organised by the IRA have a narrower anti-vax project in mind.

---

[184] There was, of course, much social media influence originating in Western countries.

[185] Collins et al. (2019).

PART III

Consequences:
Science, Truth, Democracy and the Nature of Society

CHAPTER 10

Some Immediate Consequences of the Coronavirus (COVID-19) Pandemic for Science

In the Introduction we signalled that the coronavirus (COVID-19) pandemic, which provides a living experiment bearing on many of our theses, would reappear throughout the book. This chapter is entirely devoted to some of its consequences. One consequence is a febrile enthusiasm for the demise of the scientific conference circuit triggered by the success of the remote communication enforced by COVID-19 lockdowns. Obviously, such an enthusiasm flies in the face of the central argument of this book, which turns on the importance of face-to-face communication in the generation of trust with scientific conferences and meetings being the principal example. Therefore, because the way science uses face-to-face communication in its search for truth is a foundational part of our entire argument, we now argue that the abandonment of the face to face is exactly what should not be done in science, however appealing it seems to be in the light of the success of remote platforms under the pandemic. If science is destroyed it can no longer act as check and balance or object lesson for decision-making.[186]

A classification of face-to-face encounters

Right at the centre of the analysis is the idea that all institutions that value integrity will make decisions through establishing trust relations via face-to-face interaction. This is not a sufficient condition for making good decisions, but it is a necessary one. To repeat our central example, how this works is science,

---

[186] And see note 5.

which we characterise as 'craftwork with integrity.' Among the immediate consequences of the argument is that the movement to mitigate global warming and reduce inequality in science by abandoning face-to-face meetings is misplaced: though reducing science's carbon footprint and reducing inegalitarianism in science are both good and achievable ends, to achieve them by abandoning all face-to-face international meetings will destroy the crucial quality of science that separates it from opinion formation via social media. We believe that, when it is safe to do so, the international gatherings that characterise science should begin again, though in a modified form that takes account of both the climate and the inegalitarianism concerns: such meetings should be more egalitarian, and they should be justified by their scientific purpose. Where the purpose is a trade fair, the generation of funds or publicity, there is a case for cutting down, but the face-to-face personal interaction feature of valuable and well-run meetings should not be replaced by imitation face to face; imitation face to face is enormously valuable but cannot replace the trust and culture forming roles of genuine face-to-face meetings.

Perhaps one way of concentrating the mind on this argument is to consider the way that different kinds of face-to-face meetings work. Thus, the beginning of the end of lockdown in the UK, in the form of the first return of children to schools, represented only a partial return to face to face because the children had to maintain social distancing, sometimes enforced by physical barriers, had to accept constraints on group work and seating, and had to use face masks; these hide facial expression that is so important a part of the body language and is part of what makes full social interaction so valuable: there was nothing fake about that mode of the return to schools, it was *partial* face to face not pretending to be anything other than it was. Now compare this watered-down return to school with the meetings described at the beginning of Chapter 7, both boat trips, one on Southampton Water and one a river trip. These kinds of river trip go beyond the far end of the spectrum of normal face-to-face meetings, exaggerating the positive features. In both cases, the commitment and effort to engage, which we have argued, is energizing in respect of the goals – the very inconvenience is a positive feature – was more than normal: if you wanted to be there you had to get on a boat and stay on a boat. And there was no escape: you had to be there talking to the other participants in an exaggeratedly small space – there was no sloping off – the participants were disconnected from the outside world. That both trips had very successful outcomes in terms of consensus building is almost certainly no coincidence. We can call this *enhanced* face to face. There is also a hyper-committed version intended not primarily for *consensus* building but *team* building, group solidarity being the aim rather than cognitive agreement. This is the goal in army training or in management team building where stressful physical effort is central. Beyond even this are cults with permanent gatherings in purpose-designed locations, where the intention is to re-socialize a group permanently. And even beyond this is basic primary socialization. We have, then, described seven categories of face-to-face engagement; set out in decreasing order of intensity these are *primary; cultish; team building; enhanced; regular; partial* and *imitation*.

A nice thing about this approach, and what follows from it, is that it makes better sense of the fractal model of society. All human societies start with the intensity of primary socialization, but the socialization that gives rise to the expertises in the mini-societies that are embedded within them in the fractal model is not as intense as primary socialization, but is pretty intense nevertheless. Cults, or course, have to be really intense because they aim to displace primary socialization with something fundamentally different, whereas the mini-societies in the fractal model are mostly adding something on top of primary socialization. And the same goes for the differences between the length of anthropologists' and ethnographers'/sociologists' fieldwork encounters, as described on the first chapter just after we first encounter The fractal model of society; anthropologists need intense and extended fieldwork whereas ethnographers get by with much less; anthropologists have to come close to acquiring a new primary socialization; ethnographers and sociologists are adding something on top of what they already embody.

Table 1.1, with which we started the book, can, perhaps, be used to describe these categories in a little more detail albeit this is rough and experimental – essentially a provocation to develop the approach further; we will also engage in some pseudo-quantification, not because there is an arithmetically exact description of any of this but because it directs the arguments to the right level of detail rather than allowing them to be a matter of one general claim versus another. The 12 features of the face to face, taken from Table 1.1, are listed in Table 10.1 and here each is assigned a score of 0–10 depending on their salience and intensity in different kinds of socialization (just as one might light-heartedly judge a restaurant meal on a scale of 1–10). Scores for each feature are listed for the seven categories of face-to-face encounters which have just been described. The resulting table is 'normalised' by giving each item a score of 10 for the basic socialization category and a score of five for the regular scientific meeting category – a score which would apply to innovatory business meetings and other knowledge generating social interactions which have integrity. A dash is meant to signify 'no answer' – 'that feature is not relevant here'. We also allow ourselves the indulgence of giving a score of '11' for activities which go beyond the normal.[187]

To repeat, this is not meant to be the end of an argument but the beginning. For example, the 0s in the final column could well be replaced by 1s because meeting up remotely is certainly better than not meeting up at all (crab sticks are quite tasty!). But we have left them as 0s because, *in the long term*, meeting up remotely without other kinds of meeting is likely to strengthen the illusion

---

[187] Many readers will recognise 'turn it up to 11' from the film *This is Spinal Tap*; the innovatory 11th point on the volume control is claimed to make the rock-band's amplifier still louder.

Table 10.1: Classifying the face to face.

SEVEN TYPES OF F2F ENGAGEMENT / TWELVE FEATURES OF F2F	PRIMARY	CULTISH	TEAM BLDG	ENHANCED	REGULAR	PARTIAL	IMITATION
1 Commitment and injection of energy	10	11	11	8	5	5	1
2 Learning from the bath of words	10	11	-	5	5	4	2
3 Acquiring reliance	10	11	-	5	5	4	0
4 Tacit Knowledge transfer	10	11	-	7	5	4	0
5 Learning how to trust	10	-	10	7	5	4	0
6 Domain discrimination	10	-	-	7	5	4	0
7 Immediate influence on interpretation	10	11	-	7	5	4	0
8 Body language modifying meaning	10	11	-	7	5	3	0
9 Safer adversarial dialogue	10	-	-	7	5	4	0
10 Efficiency in conversational turn taking	10	-	-	7	5	4	0
11 Efficiency in number of meetings	10	11	-	-	5	5	0
12 Serendipitous meetings	10	-	-	-	5	3	0

of intimacy and encourage the liquification of knowledge. Enthusiasts for technological solutions, or those who do not like logical extremes, will still want to argue that 0 is the wrong number in many of the rows, but the table will at least encourage them to argue about each feature in detail rather than simply describing the final column as different form of face to face.[188] For example, we can point out that, though a slight commitment of time is required to engage in a Zoom meeting or equivalent, it is nothing like travelling to a meeting; there can be no commensality or bumping into someone in the lunch queue; there can be no refusing to go out to lunch because demonstrating how to measure the Q of sapphire is more important. There can be no equivalent commitment to the project that comes from the concentration brought about by being in another place because with Zoom and the like, you can go to the kitchen and get a cup of coffee, or return to work of another kind, any time you feel like it. No doubt there could be similar discussions and disagreements throughout the table; this is what we would like to encourage. In respect of science, one can see that if one wants to try to resolve, or at least clarify, a serious and deeply felt dispute, then enhanced meetings of the small boat-trip type might be the best, but there might be other areas of science, where the taken-for-granted culture is well-sedimented and there are few disagreements, where even the remote activities of the final column might be good enough (e.g. see Appendix 4). This is the kind of analysis we need to improve the conference circuit.

---

[188] Remember that feature 9 – 'Safer adversarial dialogue' – does not mean trolling or flaming anonymously or safe from physical retaliation, but arguing strongly *without* falling out, which is of value in knowledge-building communities.

Coronavirus (COVID-19) and scientific conferences: what should *not* be done!

To return to what should be done, in Chapter 2 the results of interviews with scientists concerning their reasons for making expensive and time-consuming journeys to scientific conferences and meetings were presented. The lockdown following the Covid pandemic gave rise to a 'natural experiment' on the use of national and international scientific gatherings. Suddenly they were no longer possible. Many scientists grasped the new opportunities presented with enthusiasm. Always worried about science's carbon footprint, they realized that much of what traditionally took place at meetings could be managed via platforms such as Zoom. As time went on, anecdotal evidence suggested that users of Zoom or equivalent realized that something was missing in the liveliness of debate and the increase in fatigue consequent on the absence of the 'energy of the crowd'. But these seemed a small price to pay for the cancelling of conference-related air travel's contribution to 'saving the planet' and, a newer concern, the way the traditional 'conference circuit' reinforced elites and disadvantaged early career researchers, women and minorities. Given the thesis of this book, however, too much enthusiasm for such a change in the way the business of most, or at least many, kinds of science is conducted ought to be viewed with concern.

The 'natural experiment' provided the opportunity to conduct a small piece of survey research which reaffirmed many of the claims made in Chapter 2, but this time across the field of photonics in addition to gravitational wave physics. The questionnaire also gathered some information on frequency of conference attendance and new working partnerships established. The first report on the questionnaire study is accessible online.[189]

The hypothesised impact of the shutting down of scientific conferences, according to the analysis of this book, is illustrated in Figure 10.1 – it increases with time.[190] The front plane of the three-dimensional space represented in Figure 10.1 is the moment of lockdown when F2F ceases and is replaced by R2R, while the effects unfold over time as represented by the Z-axis going into the page. The X-axis is the extent to which a science is radically novel and/or beset by controversy of the sort that is best resolved by face-to-face debate and where consensus building coextensive with the growth of new trusting relationships is important, as opposed to settled and routine, at least in the short term and more likely to work with an information exchange model.

In current work we are exploring the nature of the X-axis more deeply; as things stand, the most determined enthusiasts for a shift to remote working do

---

[189] Collins et al. (forthcoming 2022).

[190] A short (one-page) version of this argument (without the figure) was published as Collins, Barnes and Sapienza 2020 in the journal *Physics World*.

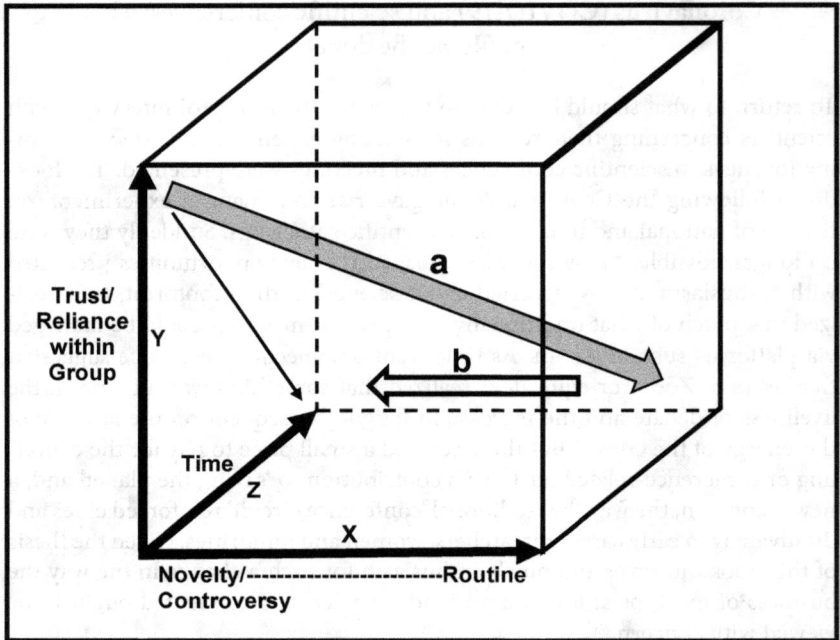

Figure 10.1: Replacement of F2F with R2R over time.

seem to come from the medical and life sciences, and it may be that a higher proportion of their work involves the application of well-established procedures rather than the continual presence of the kind of doubts illustrated by the story of the Q of sapphire and the reasons for attending conferences expressed in Chapter 2. There is a lot of use of automation in molecular biology and a great enthusiasm for more of it, it sometimes being believed that it could solve the replication crisis and other major conundrums in the philosophy of science.[191]

The Y-axis represents the extent to which the domain is already character-ised by dense taken-for-granted and trusting relationships at time zero. We expected, based on this model, that the initial impact of the lockdown would not be severely felt once the technical means for video-conferencing had become routine.

The likely consequences of the shift to R2R as they unfold over time are rep-resented by the thin arrow on the left face and the arrows labelled 'a' and 'b'.

---

[191] See Collins et al. (forthcoming 2023), for a description of molecular biology which describes it as a 'hyper-normal science' and locates it at the far-right end of the X axis of Figure 10.1; the paper tries to explain what is special about molecular biology. The papers by Alkhateeb (2017) and Wykstra (2016) display an enthusiasm for automation that reflects ignorance of the well-understood problems of replication in less routinised sciences.

The thin arrow indicates reduction in trust and reliance as the impact of the replacement of F2F by R2R is felt over time. This diminution of trust and reliance can be represented on the two-dimensions only of the left-hand plane of the figure. At time zero routine sciences would be located on the right of the space, possibly at the bottom right-hand corner and could be expected to be relatively unaffected by the passage of time unless or until they encountered a potential radical change in their taken for granted world.

Novel/controversial sciences start at the top left-hand corner of the figure, requiring dense trust and reliance relationships to work reliably. Arrow 'a' is a possible first response of a such a science to the hypothesised diminution of trust and reliance that accompanies the shift from F2F to R2R and becomes more marked over time. Such a science could respond to the change by becoming less adventurous over time, coming to rely more on established knowledge and falling toward the back right corner of the diagram; the balance of the 'essential tension' that characterises such a science will shift, 'normal science' and authority becoming more dominant and major conceptual innovations becoming less valued. A transactional logic would replace trust relations and communication would become increasingly a matter of information exchange.

But a retreat into a safer region of science is not the only possible response to loss of trust. Arrow 'b' is a possible reaction to an increase in safety which reverses the changed balance of the essential tension, such that radical innovation becomes a dominant value as is found in fringe sciences.[192] As we know, fringe beliefs are much more readily supported by social media and other internet-based interchanges than the dense face-to-face groups typical of science. Of course, an adventurous science might move directly to a fringe-like position without any intervening move to routine.

Elitism and the conference circuit

The perception that scientific meetings foster elite networks is another strong driver of the enthusiasm to abolish the conference circuit permanently. Therefore, we will discuss this briefly here, since it has been triggered by the coronavirus (COVID-19) lockdown, even though it does not fit so clearly with the main theme of the book, which is communication.

It seems likely that some well-known scientists gain huge prestige from the conference circuit, becoming 'famous for being famous' while attending large numbers of meetings per year (anecdotally, up to 50 on a regular basis), which cannot be justified by the kinds of reason discussed in Chapter 2; we can be pretty sure that no deep purpose related to scientific *knowledge* is being served even if the conference *business* is being served by this kind of thing and science

---

[192] That a balance shift toward novelty is a characteristic of fringe science is shown in Collins, Bartlett and Reyes-Galindo (2017).

is getting some publicity or even reinforcing the cognitive taken-for-granted in certain scientific domains.

It is also the case that the way the conference circuit is arranged is far from optimum in terms of egalitarianism: for instance, childcare facilities need to be made compulsory along with bursaries for junior researchers and those from developing countries, with more emphasis on setting aside some part of large meetings for formalized tutoring and senior-junior meeting sessions. So, there is a great deal to be put right, but the idea that one can have a science devoid of any of the features of an elite institution is based on misunderstandings of how adventurous science works, not only a misunderstanding of the importance of the face to face in generating knowledge, but also a misunderstanding of the role of communication among elites in science.

A science in which each contribution was judged entirely by its contents and never by the contributor's experience, or record of achievement, or embedding in existing trust relations, would be a science that is mechanizable; scientists could be replaced by machines programmed to carry out fixed procedures or invent new ideas and processes by randomly exploring appropriate physical and mathematical spaces. The dream is an old one: it was claimed that Langley and Simon's 'General Problem Solver', which goes back to the 1990s, redis-covered Kepler's Laws of planetary motion; in more recent times, 'intelligent' computers are being brought in to execute various routine elements of certain sciences. If this was how science could operate then human scientists working remotely would be able to do their work and feed the results into some common 'metascientific fact grinder', each contributor's work being assessed without ref-erence to certifying institutions and networks: indeed, there would be no rec-ognizable scientific expertise, only a series of instant atoms of accomplishment, mapping on to an ideal logical positivism with good arising out of the a-social activity of monads, as it does in standard economic models of free enterprise and idealized, bottom-up models of democracy.

But this model of science has not been viable outside of narrow domains since it was realized that science was a cultural activity, not an 'algorithmi-cal' activity. The details emerged in the early 1970s, with the development of the sociology of scientific knowledge, but its roots go back to the 1930s with Ludwik Fleck's *Genesis and Development of a Scientific Fact* and Thomas Kuhn's idea of paradigms and 'the essential tension', first expressed in 1959:

> ... Some divergence characterizes all scientific work, and gigantic diver-gences lie at the core of the most significant episodes in scientific devel-opment. But both my own experiences in scientific research and my reading of the history of the sciences lead me to wonder whether flex-ibility and open-mindedness have not been too exclusively emphasized as the characteristics requisite for basic research. I shall therefore sug-gest below that something like 'convergent thinking' is just as essential to scientific advances as is divergent. Since these two modes of thought

are inevitably in conflict, it will follow that the ability to support a tension that can occasionally become almost unbearable is one of the prime
requisites for the very best sort of scientific research.[193]

Science has to resolve this tension between creativity and authority, and it cannot be resolved by a formula. The resolution of this tension is what locates and
maintains the boundary between the inside of science and the outside – the
fringe plus the public. Without the maintenance of such a boundary, science
would become, at best, a maelstrom of competing claims and, at worst, an institution indistinguishable from social media where it is simply the aggregation
of 'likes' that count.[194]

It is important to understand, then, that elite groups supported by face-to-
face meetings cannot be eliminated if science is to be distinguished from opinion. The job of those groups is to reach conclusions about what is a legitimate
part of developing scientific culture and what is not, and therefore *who* is to be
treated as an insider and who an outsider. The focus should not be on eliminating these meetings but on making sure that they work in a way that adheres to
the Mertonian norm of universality. That is, they should work to include and
exclude, as far as possible, according to scientific accomplishment and earned
trust, not the characteristics of the scientists such as gender, age, race, sexuality,
nationality or physical ability.

This is not to say that organizing meetings that rectify the wrong kind inegalitarian biases is easy however conscious conference organizers are of what
needs to be done, and however assiduous they are in doing it. This is because
the fundamental sources of inegalitarianism arise from structural features
of the profession of science *as a whole*. Most obvious are the almost inescap-
able educational disadvantages of relative material poverty, whether within or
between nations, often correlated with variations in the salience and acclaim of
educational achievement in different cultural groups, and, of course, the career
pressures experienced by caregivers of all kinds.[195] But these structural problems are not the problem we are dealing with here: we are asking about what

---

[193] First mentioned in Chapter 3: Completing the Story of Face-to-Face Communication 3 and first referenced there in note 93 (Kuhn 1959/1977).

[194] For an analysis of the fringe see Collins, Bartlett and Galindo (2017) and
note that not even scientific qualifications can distinguish an inside from
an outside since the fringe is populated by highly qualified scientists often
working in universities.

[195] In addition, Delamont (1989) argued that women can be attracted to
science precisely because they see it as meritocratic rather than relationship-
building. She shows that even when attending conferences women can easily miss some of the key benefits that turn on trust-building activities: they
are less likely to attend conference meals or participate in conversations at
the bar and may well see these as unimportant, unenjoyable or indeed risky.

should happen to, and at, scientific conferences and meetings, and arguing against the notion that the abolition of inegalitarianism at conferences – which is something that should be achieved with urgency – will abolish inegalitarianism in science.

Worse, abolishing inegalitarianism in conferences by abolishing conferences themselves could enhance inegalitarianism in science as a whole, even as it damages science beyond repair. In the past, those with caregiver responsibilities have experienced conference attendance as welcome havens during which they could work uninterruptedly, safe places that have been lost under the lockdown.[196] The other danger is the temptation to create a disguised two-tier system – participation for those who can afford to travel while 'equal quality participation' is advertised through accessible remote link. This could create something like the two-tier educational system found in the UK, where top universities' recruitment is heavily biased toward applicants from private schools rather than the state system even though, on the face of it, the two systems provide an equal educational experience. That is another reason to avoid the growing tendency to re-label remote video interaction when it takes place on platforms that show participants' faces as 'face to face', as though it were a proper substitute, when it has little of the richness of personal interaction; this, as suggested, is often a kind of subtle advertising for the remote.

The obvious answer to the development of two tiers of conference attendance, one involving F2F and the other imitation F2F, is complete abolition of travel to meetings, so all would be imitation. But even that, disastrous for science though it would be, would not eliminate a more subtle version of the two tiers: as it is, those student-scientists, and other scientists, located at major universities such as Harvard, MIT, Oxford, Cambridge and others in the top rank, however one cares to define it, along with institutions physically close to them, already have easy interpersonal contact with the top professionals, whereas those in minor or distant institutions do not. This much more undesirable elitist 'old-boy network' would not be eliminated by abolition of the conference circuit; instead it would be disguised and enhanced, since conferences would no longer provide a venue for those without significant scientific cultural capital to meet those with it.

In sum, the desire to reject the conference circuit on the grounds that it is elitist is misplaced because elitism is a necessary part of science, but the trust involved in establishing the right kind of elites requires the face to face; to get

---

[196] This article illustrates the problem for women researchers with children: https://www.universityaffairs.ca/news/news-article/women-academics -worry-the-pandemic-is-squeezing-their-research-productivity/. The advent of the word processor and the internet has made a huge change in the way overseas conferences are experienced for everyone: in the days of the typewriter they were time taken away from work, nowadays they are an opportunity for long periods of uninterrupted work.

rid of scientific meetings because they do not work in an optimum way is to throw the baby out with the bathwater. The criticism should be aimed at making science not less elitist, but less inegalitarian – the two things are not the same: 'elitist' means having a well-maintained inside and outside, based on trust and achievement; 'inegalitarian' means that only people with certain ascribed characteristics can get in; the former is necessary, the second genuinely undesirable.

To repeat, this apparent diversion into the narrow matter of conference attendance is actually not a diversion from the main theme: democracy needs science and science, if it is not to collapse under the pressure of social media and the like, must maintain trusting face-to-face relations among its core-set members for all the reasons put together in Part I of the book. Science and its way of being are central to the survival of democracy both as a check and balance and as an object lesson in decision-making.

CHAPTER 11

The Nature of Democracy and Scientific Expertise

Moving to wider issues, the problem is how to preserve and nurture a truth-based culture that seems ever more under threat in the US, UK and certain other 'Western' democracies. As we intimated in the opening paragraphs, it is not impossible to imagine a future scenario in which people will look back and say that though communism lost the Cold War with the fall of the Berlin Wall, it was always going to be a matter of 'mutually assured destruction' ('MAD' as the prospect for all-out thermonuclear was termed in the 1960s). As we will suggest, the apparent victory of capitalism in 1989 created the conditions for such an unrestrained championing of belief in freedom of action and expression under free-market capitalism that it had a good chance of leading, eventually, to the self-inflicted demise of democracy too. And now we have a new technology, the internet, that, with very few resources and no bullets or bombs, could allow both us and others to bring this about sooner rather than later.

To repeat, too-readily accepting remote communication as a substitute for face-to-face communication will facilitate the transformation we are trying to avoid. In Part II we argued that even in cases where remote communication is of enormous value, the underlying trust that supports all sound communication is still grounded in the face to face. As we explain in the Introduction, Part III links the problem to the nature of democracy; in this chapter we explore this relationship further. The overall argument builds on analyses in previous works suggesting that science can offer leadership in respect of many of the values that support democracy in general, and that scientific expertise is one of the checks and balances that support 'pluralist democracy'. In the most recent of these works, we define populism by contrasting it with pluralist democracy.[197] Now, we want to set the idea of pluralist democracy into a more general

---

[197] Collins and Evans (2017b) is entitled *Why Democracies Need Science* and argues that science could offer leadership in respect of many of the values

How to cite this book chapter:
Collins, H., et al. 2022. *The Face-to-Face Principle: Science, Trust, Democracy and the Internet*. Pp. 177–187. Cardiff: Cardiff University Press. DOI: https://doi.org/10.18573/book7.l. Licence: CC-BY-NC-ND 4.0

framework by dividing democracy into two simple classes encompassing the many standard subdivisions (see notes 10 and 202) in a way which resolves some confusions and misplaced ambitions about the role of science within it. The final chapter discusses what we should be trying to do next; these simple models will pave the way.

'Popular assertive' and 'structured choice' democracy

What we will call *popular assertive democracy* (PAD), defines the essential feature of democracy as decision-making by the people; it works as one might say, from the bottom up: the people are to have the decision-making rights and where they do not, something has gone wrong. This model goes back to the Greek City State and is reaffirmed by political thinkers like Rousseau. For those who support this model, to show proper respect for the citizenry is to continually assert and allow them to exercise their rights in every phase of politics and policymaking. The model conceives of the citizen as born with inalienable rights and treats full citizens as having, not only the right, but the ability to make free and unconstrained choices.

In contrast, for what we will call *structured choice democracy* (SCD), the defining feature is the distribution of power amongst the various institutions of society, the population as a whole being one of those institutions; SCD treats the citizen as largely a creature of society, as conceptualized by the sociology of knowledge; animals are, in this sense, not social beings: they are created by their evolution not by their language and culture – humans are created by their language and socialization. This does not mean that isolated human individuals do not have choices to make, the kinds of choice that put one party in power rather than another, but the range of those choices is narrow. The organic nature of society remains fixed and provides the envelope of choices such as are enumerated at election time. A few humans manage to break away from their early socialization by special effort or in consequence of special circumstances, but the starting point for humans is embedding in the societies that make them what they are, as described in the first divisions of Table 1.1 and further illustrated in the fractal model of society, Figure 1.2. SCD is not something we choose or do not choose: it is a scientific description of human life in a democracy; the only choice is how we describe or misdescribe the nature of that life. Human life is misdescribed when it is said that we are ever completely free to make choices; our freedom is always limited to a restricted envelope

---

that are central to democracy; Collins et al. (2019) is entitled *Experts and the Will of the People: Society, Populism and Science* and introduces the term 'pluralist democracy'. The meaning of populism in this treatment is similar to that of Jan-Werner Muller (2017), and its relationship to established knowledge is developed in Collins et al. (2019).

provided by our socialization into different groups which also provide our varied abilities. The native language we speak is an iconic example of the point – we are provided with a language through our socialization, and we are limited to that language – along with any others we can acquire through exceptional effort in later life.[198]

Given that we live in an SCD, there is a tension between society's institutions and groups, each inhabited and formed by citizens who have undergone different types of socialization and acquired different experiences and skills; the tension is not between individuals but between institutions and groups. Institutions and groups, as shown in the fractal model of society, include the government, its various advisory and constraining bodies, and the adult population as a whole, who have acquired the ubiquitous expertise that makes them experts in the nature of their national society and its major sub-groups. Once we recognise that we live in an SCD, it becomes clear that none of these institutions and groups should dominate decision-making. When a certain set of citizens distributed among these institutions and groups vote in such a way that their choices add up to the election of a particular administration, this does not change the structure of society nor the rights of citizens within its groups, even though populist leaders will claim it does.[199] Recognizing the variation in the envelope of choices across the fractal model of society – the expertises of different institutions and groups of citizens – no longer treats each individual as a free-floating decision maker but as a representative of the set of groups to which they belong. This is illustrated in Figure 11.1.

Figure 11.1 does not represent a political position; it represents a social scientific description of the reality of human life under a democracy. Even political

---

[198] Friedrich Hayek, who, in 1944, published the iconic book justifying free markets and neoliberalism, *The Road to Serfdom*, argues that a planned economy can never cope with the variety of preferences of citizens, (e.g. pp. 62–3), and this has, perhaps, been misinterpreted as an ideology of free choice. But Hayek does not give extended consideration to the sociological source of our choices nor the limits set by our socialization and competencies. Oddly, Hayek's own premise about the varying choices of individuals is qualified by his remarks about propaganda as a controller of thought, found in his Chapter 11. Indeed, since in that chapter he says the danger of totalitarianism is that there is only a single source for the views of citizens, his stance is that citizens in a democracy have varied sources of views – remarkably like the model represented in Figure 11.1! He even goes on to admit that only a few citizens can break away from the thoughts given to them by society (p. 68 ff.). His argument about the superiority of markets over centralised planning may be right in many respects but this does not imply that humans are free as individual actors.

[199] See Collins, Evans and Durant (2019, Ch. 2) for a longer analysis of the relationship between the enumerative and organic faces of society.

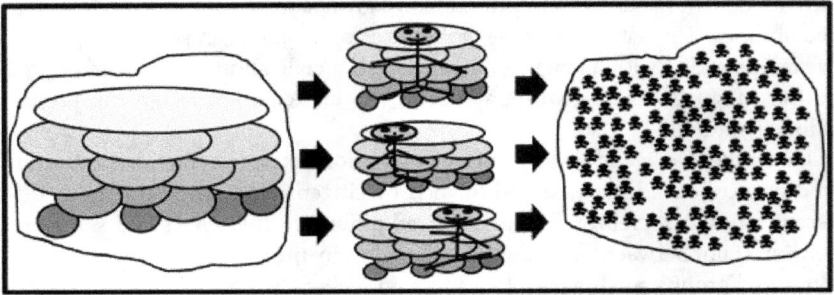

Figure 11.1: *Popular assertive democracy* (right) is disguised *structured choice democracy* because each supposedly free and independent individual is an aggregate of the social groups which enable their abilities and choices; in effect it is not free-floating individuals but the groups which are choosing, *via* their representative individuals.

right-wingers such as Andrew Breitbart recognise that culture comes before politics, though he has a shallower view of culture. The deeper social scientific truth does have an impact on how we should understand the nature of democracy which is different to Breitbart's.

This model offers proper respect to the citizenry and the society they create, and that creates them, by recognising the way certain citizens strive to acquire different expertises so as to build and serve a complex society. This model constrains the power of any one institution or group including the government; the government is constrained by, among other things, bodies of experts, while the experts must be constrained by the government in such a way that the society does not become a technocracy.[200] The constraint on the government is what separates this kind of democracy from populism, and the earlier term, 'pluralist democracy', captures this feature well. It is a good term because it stresses the way power is distributed across groups with different interests and opinions rather than concentrating it in the majority that emerges at election

---

[200] Darrin Durant explains that one way of describing the major types of democracy, the approach being due to V-dem (https://www.v-dem.net/en/) is by dividing them into five: participatory, deliberative and egalitarian, these three being concerned with how the public makes its decisions, electoral and liberal, where electoral points to all the structural relations that a democracy needs to be a democracy, such as free elections and freedom of association and rights, while liberal points to the principles and norms that govern how to flesh out the structures, such as how to constrain power relations.' He suggests that the popular assertive model encompasses the first three while structured choice democracy covers the other two (private communication, 9 October 2020).

time, so we will continue to use this positive term whenever it seems appropriate, remembering it is a subset of the SCD that is one side of the more general division set out here.

SCD includes a focus on what happens at the top of society – the government – so there is an element of constraint on what happens 'top-down', which is, in a sense, the complement of the attention given to the 'bottom-up' choice in PAD. There is a moment at which top-down constraint and assertion of bottom-up power converge to a common focus; this is because one of the constraining mechanisms on the power of governments, and the institutions to which they distribute power, is accountability to the people as a whole via regular general elections and this, of course, is an essential feature of PAD too.[201] We will argue that moments like this aside, SCD is a consistent position whereas PAD is beset with paradoxes, dangers and, as we will see, when it comes to the role of expertise, magical thinking.

Populism fits squarely with bottom-up PAD and runs inextricably foul of SCD. Populism resolves all the paradoxes of assertive democracy by locating all decision-making power in the hands of a subset of citizens, usually a majority group, and treats their views as constituting the 'will of the people'; it then transfers that power to that group's representative, the government.[202] Any opposition is treated as traitorous as it opposes the will of the people. Bottom-up democracy and populism are natural bedfellows, as some political analysts readily understand.[203]

Populism is, of course, incompatible with technical expertise and so it attacks experts as traitors whenever they in conflict with 'the will of the people'. Pluralist democracy, and SCD in general, treat the losers of an election as still having rights and their views as still having potential value. The constraint on the power of the majority and on the leadership is maintained via 'checks and

---

[201] Durant (private communication) also points out that there is also a common focus on providing active 'protections' for certain freedoms (though this tends to be over-interpreted in capitalist societies, leading to distrust of the state). The extent to which this electoral 'moment' ought to be extended to other forms of interaction between elections, such as popular protests, is a matter of debate; it might also be the case that continual civic debate is part of the socialization process that leads an electorate to maintain an understanding of what they mean by democracy so that voting is an informed, not a momentary choice. See also Durant (2018, 2019).

[202] Note that in the election of President Trump and the one that reaffirmed Brexit as a result of the electoral college system in the US and political parties in the UK failing to form alliances, it was a *minority* of the population whose preferences were counted as the will of the people.

[203] E.g. Mouffe (2000). But see also Bonikowski (2017) who indicates that populism, in spite of the rhetoric of the will of the people, also depends on the attraction of ethno-nationalism and similar appeals.

balances', including a 'loyal opposition', alternative chambers, a judicial system and a free press, which prevent the ruling party's views from completely overwhelming the views of the opposition and other dissenters.[204] Scientific and technological expertise is also a check and balance and that is why, and this should now be obvious from both the arguments in this book and recent events, if SCD is to be preserved, a high valuation for scientific and technological expertise is necessary, just as placing a high value on judicial expertise and journalistic expertise is necessary.[205]

The role of the state in democracy: positive and negative freedom

One knotty problem found in standard treatments of democracy, carefully and deliberately sidestepped by the terms 'pluralist democracy' and 'SCD', is the tension between positive and negative freedom. Positive freedom stresses the need for individuals to be enabled to act; negative freedom stresses the removal of interference over free choice.[206] A problem arises when the political and economic defeat of communism is taken to represent such a decisive triumph for the idea of negative freedom that this, along with unconstrained free markets, leads to power being relocated in ways that will destroy democracy just as surely as a military or economic victory for communism would have destroyed it. We can see the tendency in right-wing rhetoric favouring negative freedom and free markets: in the UK, the idea of positive freedom is ridiculed in the term 'the nanny state'. The 'nanny state' rhetoric takes it that citizens are adults and know how to choose and order their lives using their common sense without the need for a child's nurse. The rhetoric downplays the role of the state, fulfilling the same discursive space as, in the USA, the evil 'big government'.

This overenthusiasm for negative freedom also rests on magical thinking: the idea that individuals' unconstrained action will lead to the greatest utility for all. One can just about understand how there could be an argument in favour of the idea when economic markets are the focus, but not when we turn to other kinds of freedom. But even economic markets do not work without regulation, as critics of standard economic theory point out; we saw the fallacy starkly exposed in the 2008 banking crisis discussed in Chapter 6. Furthermore, the

---

[204] Recent events in the UK concerning Brexit illustrate these points with almost embarrassing faithfulness.

[205] Trump's attacks on climate change, and the fictitious claims about the size of the crowd at his inauguration, are also almost too obvious illustration of the point. The notion of alternative facts and the liquification of knowledge make the world safe for populism as defined above.

[206] For an account of the way negative freedom in modern American culture is understood, see Thorpe (2016), Ch. 5.

great champion of free enterprise, the United States, *does not act* as though free markets are the wellspring of riches for all, or even riches for some.[207]

Quite apart from control over monopolies, pollution and the like, the so-called free-market capitalism, which led to the immense wealth of the aerospace industry, the internet and 'Silicon Valley', was actually heavily state sponsored through the creation of decentralised public innovation agencies that invest not just in high-risk venture-capital type funding but right across the innovation chain. These include the National Institutes of Health (NIH), NASA, the Defense Advanced Research Projects Agency (DARPA), its sister organization in the department of energy (ARPA-E), the Small Business Innovation Research (SBIR) programme and the National Science Foundation (NSF).[208] As an example, 'entrepreneurs like Elon Musk have received guaranteed loans from the US Department of Energy, with the LA Times estimating that his three companies (Tesla, Space X and Solar City) have together received around $5 billion in public support.'[209]

When we turn to actions which are not primarily economic, the state helps in two ways. First, there are regulations that allow citizens to live in the first place, and to live in safety and security, through the enactment and sustaining of laws, regulations, enforceable standards and public awareness. That is why the glass in shower-screens is shatter-proof and the sleepwear worn by children is flame-retardant; it is why (nearly) everyone drives on the same side of the road at reasonable speed and wearing seatbelts; it is why there are regulations about smoking in public places and discharging sewage outside of state-provided sewage systems, clean water in the taps and regulation of food additives. The idea that individuals could spontaneously manage this without a state to provide the organizing structure is a fantasy (as Hayek saw clearly).

But then there are the other functions, not of the state, but of the society, that are alluded to in the first section of Table 1.1. The society enables people to live in the first place – it provides the conditions for socialization without which we would be simply human animals equipped with little more than the instincts needed to survive brutishly and reproduce. The society provides us with a native language and a set of templates for thought and action that is, or was, distinctive to each nation. The myth holds that at some point humans break away from their nanny and become free. But we never break away – just look at the differences between what citizens take for granted in Britain and America, two countries which, to plagiarise, are divided by a common culture.

---

[207] Hayek (1944) set out the essential limits on freedom needed for economic markets to function (see, for example, pp. 39–41).

[208] See Block and Keller (2011) and Mazzucato (2013). (Thanks to Josh Ryan-Collins for help with this section on US government support for the 'free market.')

[209] Mazzucato (2017), referring to Hirsch (2015).

Nearly all UK citizens find the love of guns and the adherence to fundamentalist religion, that are widespread in America, deeply alien.

And culture is always in need of maintenance and always developing: every act we do is a contribution to the constraining and enabling culture in which we live. Therefore we, and the governments we elect, cannot avoid the responsibility of shaping that culture. And this is becoming ever more in need of recognition as the speed of cultural change increases in response to the spin of the internet knowledge liquidiser. Those who know how to use the liquidiser have a powerful cultural weapon. The internet enables the old structuring by local groups to be replaced by distant and disguised parties, not removing the structured choice nature of society but changing and hiding the source of influence. We will turn back to these themes in Chapter 12.

Science, knowledge and democracy

From the Greek City State to today's world, it has often been said that there is tension between democracy and expertise, notably scientific expertise. This is because esoteric expertise is elite and impenetrable, and therefore it is hard to see how ordinary citizens can participate in decisions that turn on esoteric expertise.[210] Once upon a time the problem might have been settled by the general acceptance that, this tension aside, scientific expertise was self-evidently the best kind of knowledge there was because it has a way of producing certain and unbiased conclusions. Since the 1970s, however, that is no longer clear.

Among academic sociologists and philosophers, a new solution has become popular to the new conceptual problems that emerged when science ceased to be self-evidently our most perfect form of knowledge. The solution is to turn to the people when difficult decisions must be made, demand that scientists 'show their working', even bring citizens in to co-produce scientific research. We will call this 'the democratization of science' movement, and it draws what life it has from levelling down of scientific knowledge to the point when the ordinary citizen no longer feels they must respect it automatically and that, maybe, their own point of view is a reasonable competitor (an attitude clearly at work when it comes to anti-vaccine choice).[211]

---

[210] Durant (private communication) points out that Rousseau saw democracy as involving individual citizen judgments and not any kind interaction: Rousseau cuts his citizens off from listening to each other, a model which begins to fail as the scale of society increases. Durant explains that Pitkin (1967) noted that the model was only even remotely compatible with democracy in a very small community of citizens who were also heroically public spirited.

[211] For a discussion of the sense that citizens have as much expertise as scientific experts in the light of the levelling out of knowledge types, see Collins

Demanding that citizens be brought into scientific and technological deci-sion-making is reasonable in certain 'delineated cases' such as where value judgements are involved in technological implementation, or because citizens have extensive experience in the way technologies are actually used, or have specialized local knowledge about the way scientific information is gathered or special access to scientific data, but these cases have to be thought through case by case and individually analysed, not just encouraged under the broad (but seductively appealing) banner of democratization of science. We will look at such 'delineated cases' in greater depth in Appendix 3 but, in the meantime, we will separate them from the main discussion of the relationship between ordi-nary people and technical specialists in democracy, criticizing the attraction of the undifferentiated democratization of science.

Part of the attractiveness of the broad, unspecified democratization of the science movement in academic circles is that the 'more democracy' slogan has an appeal that is hard to deny: it is hard to mount an argument that favours elit-ism. Furthermore, it is said that eroding the elitism of science through this kind of cooperative work will reduce the distrust of science and increase respect for it even in the age of the internet, the problem we are struggling to solve. But we think that this whole approach is misplaced.

'Democracy's' tension with scientific expertise, said to be a consequence of science's elitism and impenetrability, is in tension only with PAD not with SCD. The idea of scientists 'showing their working' resolves that tension for PAD but only if, in some mysterious way – this is where the charge of magical thinking arises – the staggeringly complex world of science, which is so impenetrable at the frontiers, that even professional scientists are always disagreeing about whether 'their working' is sound and how it should be interpreted, can be made transparent to the public at large.[212] Citizens are taken to have a special wis-dom beyond that of experts which makes it possible for them to resolve oth-erwise impossible philosophical problems. A jury may make sense in a legal courtroom where it is guilt and innocence that is being judged, a matter of ubiquitous expertise familiar to citizens from their day-to-life, but even court-rooms run into trouble when judges and juries must decide between the views of expert witnesses. As it is we seem to be being offered a kind of 'epistemologi-cal crowdsourcing': 'when the problems become too hard for the professionals, ask for help from the crowd.'

A driver of this magical thinking is, perhaps, the perfectly acceptable ver-sion of philosophical crowdsourcing which is found in courtroom juries and their writ-large equivalent, general elections. But, as intimated, here both types

---

(2014). For the broad terrain of the argument within science and technol-ogy studies, see Mirowski (2020).

[212] See Collins (1985/92) for disagreements among scientists. Note, in contrast to the training of citizens' juries, the amount of work that social scientists put into understanding the working of the scientists they study.

of democracy converge. In general elections and in regular court cases, members of the public are having their *ubiquitous expertises* recognised – in these situations the public is being asked to vote on things in which they *are* experts because these things are central to their everyday lives: guilt and innocence and competence in the handling of a nation's affairs (given that the citizens' ubiquitous expertise in institutions like democracy has been maintained or developed).[213] So general elections and legal juries recognise the distribution of expertise among the citizenry and recognise the ubiquitous expertises in the institution of universal franchise. And – the exception that proves the rule – even legal juries go wrong when the topic becomes too technical – which is why, in the UK, juries are not used in complex financial legal cases, and why the evidence of expert witnesses on technical matters can be so misleading to juries everywhere. It is also why referendums which involve technical matters are misplaced, a prime example being the Brexit referendum where the driving force for many participants might have been a nationalistic choice, but an important consideration was economic impact and the effect on employment; here citizens were left vulnerable to misleading information and easily swamped by carefully managed political rhetoric.[214]

SCD suffers from none of these paradoxes, nor does it need magical thinking in respect of either the wisdom of citizens nor the easy and rapid penetrability of esoteric expertise. Instead, among the institutions to which power is distributed are elite groups of experts. There is a danger here and it is that certain technically expert groups believe themselves too powerful and mistake their expert advisory role for policymaking rights. That is one place where other expert groups such as social analysts and philosophers of science can contribute as constrainers of power, both of elite technical groups and of political elites who would distort the findings of technical elites to suit their purposes.[215] And that is why the citizens have decision-making rights over these things at the time of elections – they have the right and ability to say whether or not the delegation of power to elite groups has been mishandled by the government.[216]

---

[213] The idea of ubiquitous expertise is explained in Collins and Evans (2007).

[214] Of course, citizens are called upon to make complex decisions, not least about economic policy at election time, but democracy is acknowledged to be, not perfect, but the 'least worst' way of choosing governments. In any case, at general elections citizens are not expected to make specific technical decisions but use their meta-expertise to judge how technical decisions have been managed on their behalf by different groups of politicians.

[215] Collins and Evans (2017b) suggest an institution 'The Owls', to formalise this role.

[216] Of course, there is another crisis that we face in the age of the internet and which is central to this book: whether there will ever be a way to assure ourselves that elections are fair given the power of forces both inside and

That even those academics whose avowed loyalty is to the democratization of esoteric expertise, including scientific expertise, understand the paradoxes, is clear: most of those academics abhor the 2016 election result and the way it was brought about by social media-managed sloganeering, and they abhor the subsequent actions of President Trump even though they are object lessons in the workings of PAD; most of them abhor the way the Brexit referendum was conducted. And all of them abhor the presentation of UK judges in right-wing newspapers as 'enemies of the people' when they ruled, *drawing on their elite status as interpreters of the law*, that Parliament must be consulted over the execution of Brexit, thus temporarily overturning the government's ambitions – a perfect example of SCD in action and the way the notion of elite experts is central to it. We all know, if we think about it, that we need expertise if we are to have a democracy that makes sense; expertise is one of the foundations of pluralist democracy, it is not its enemy.

---

outside a nation to influence opinion by electronic means in the ways described in earlier chapters.

CHAPTER 12

What Is to Be Done?

We have looked at one thing that, in the light of the argument of the book, ought *not* to be done: we should not close down face-to-face scientific conferences permanently following the coronavirus (COVID-19) lockdowns even though we should seize the opportunity to improve the way they run and reconsider which are essential and which are not. We have also regretted the dilution of science as craftwork with integrity that comes with its transformation by platform capitalism and with every kind of corruption (see especially note 4). Now we have a far more difficult task – to say what should be done if we want to turn back the tide of liquid knowledge that threatens to drown democracy. This is the wide end of this book's cognitive funnel.

We have to start by admitting to the limits of our power as authors. While there are differences, we are all primarily researchers not activists, and with a few exceptions we have relatively little to do with the institutions of policy-making.[217] We certainly have no access to the big levers of political change – financial and military force, state organised propaganda, mass advertising, and newspapers and other mass media – but we wish we had. All we can hope for is that our analysis will influence political and civic culture in general and, perhaps, certain individuals with power. What is sure is that if, as a society, we do not understand the world, there is no chance that powerful players will know how to improve it or even that it needs improving. Furthermore, unless people

---

[217] Among the authors, Innes's work on disinformation, first discussed in earlier chapters, does inform the UK's security forces directly; Innes is the closest to having direct power in regard to what is discussed here. This should be clear from the studies of disinformation in the above chapters and in Appendix 2. In that appendix Innes also discusses some of the concrete necessities and outcomes of this kind of work. McLevey engages in some demonstrations and lobbying as well as advising; Kennedy has also advised government policy-making committees.

How to cite this book chapter:
Collins, H., et al. 2022. *The Face-to-Face Principle: Science, Trust, Democracy and the Internet.* Pp. 189–217. Cardiff: Cardiff University Press. DOI: https://doi.org/10.18573/book7.m. Licence: CC-BY-NC-ND 4.0

talk about it, there is no chance that citizens, at election time, will know what it is they should be holding politicians to account for.

Restricting the power of the internet

There is one concrete thing that relates to communication that could be done and is being done to some extent: restricting the power of the internet to spread disinformation and misinformation.[218] The illusion of intimacy will not go away but its potential as a supers-spreader of post-truth might be limited to some extent if this can be done to some effect.

The burden of detecting and responding to disinformation needs to be based upon new forms of institutional collaborations between the commercial businesses who run the platforms, government agencies, and researchers, journalists and NGOs with the readiness and ability to spot what seem to be fraudulent accounts and fake news content. The expertises needed to manage this include technical data-science skills to understand how different platforms work, how they can be 'gamed' and how to exfiltrate data from them and analyse it; social/behavioural/political science backgrounds to afford an understanding of motives and how such materials fit within broader processes and systems, along with a degree of expertise about the authoring states/groups/movements of such material, to understand their intents and objectives, and also to be able to detect how they are signalling the deception in terms of the linguistic and visual codes they are using that might be an indicator of inauthentic behaviour. At the time of writing, however, the way responsibility is distributed is unclear and unsatisfactory.

Pretty much every major public event or issue is now a magnet for disinformation so that there are a diversity of modes and types that need to be countered. Sometimes the sources can be uncovered and such cases are at the 'simpler' end of the spectrum since, once it is know that a source is malign, someone could be directed to try to control the distribution; the difficulty is

---

[218] This is, of course, being directly helped along by Innes. What we call 'the police' was originally an institution created in response to the disruptions to social order arising from the industrial and urban revolutions (Reiner 1992). They were introduced alongside other regulatory institutions intended to manage and mitigate potential public harms. The information revolution, we are suggesting, is as profound as the industrial revolution so it is unsurprising that we should feel the need for new institutions to police the novel harms that are arising from the corresponding transformation of social reality. (On the morning of 15 November 2020, Keir Starmer, leader of the opposition Labour Party in the UK, called for emergency legislation to make internet-promulgation of anti-vaccine disinformation illegal.)

executing the solution, and currently there is little evidence that measures are being effective.[219]

More difficult is the control of disinformation subsequently promulgated, with the best of intentions, as 'misinformation', by citizens. Here any attempt to restrict distribution begins to look like censorship or a curb on freedom of information, and it is hard to work out an 'objective' method of identifying the false and dangerous. The case of the revolt against MMR is a perfect example of citizens, with the best of intentions, amplifying misinformation or disinformation and, along with some UK Conservative Party representatives, believing it to be their right to do so. Ladislav Bittman, deputy chief of the Czechoslovak intelligence service's disinformation department from 1964 to 1966, wrote in his book published in 1985 that to succeed 'every disinformation message must at least partially correspond to reality or generally accepted views' (49), and Thomas Rid's *Active Measures* details many historical examples of this strategic approach to systematic manipulation and deception. The usual aim of the source of malign information is to tweak, nudge and exacerbate already established fractures and fissures in the social order so the ordinary citizens who act as amplifiers are part of the existing political spectrum, and this means there is always going to be something plausible in the source that one would like to target. Hence citizens, perhaps supported by the ideology of negative freedom, may well resent the imposed limits, and internet companies may find restraints easy to resist. One sees how directly PAD conspires in its own destruction. Nevertheless, this has to be one avenue of approach to the problem that democratic governments must pursue, and we will mention it again in the Conclusion.[220]

The case of MMR – a 'hard case'

When it comes to the power of various proposed solutions to the problems we have outlined, we are going to make things as hard as possible by confronting ourselves with the 'hard case' of the MMR vaccine revolt. We are going to take it that readers of this book agree that it would have been better had there not been a revolt against MMR and a subsequent, and widely distributed, re-emergence

---

[219] Rid (2020) argues that it is not always harmful to the disinformation agencies to have the existence of their activities known to the target state; they can still operate, and it reduces the potential shock of a full discovery.

[220] Twitter and Facebook and some others have begun to put warning notices on, or deleting Trump's disinformation – see e.g. https://www.bbc.com/news/technology-54440662; https://finance.yahoo.com/news/twitter-label-candidate-tweets-claim-144250824.html. It is reassuring to some extent that the American news networks stopped broadcasting Trump's post-election accusations of fraud but it is hard to know how much this was just a matter of people spotting the moment to abandon a loser.

of measles epidemics, whereas before the revolt measles was very rare indeed. (Measles, by general agreement, is a dangerous and potentially maiming or even fatal disease, especially where those with already compromised health are concerned; single-shot measles vaccines are more expensive and less likely to be properly administered than MMR). We will use the MMR case as a probe to explore how citizens and others recognise and choose between mainstream scientific core sets, fringe science groups, cults and conspiracy-theory-driven groups, and how things could be improved; in turn the answer will reflect on the far larger question of the preservation of SCD with its dependence of tension between groups of experts and the powerful. We are, as it were, inserting a mini-reverse funnel into the cognitive argument: it runs from a detailed analysis of the MMR case to the wide problem of democracy.

A vaccination-related case is also pertinent because vaccination bears on individual choice versus collective good, since all vaccinations carry small risks whereas the establishment of herd immunity requires high vaccination rates, so 'the tragedy of the commons' and 'prisoner's dilemma' choices are at the centre of the discussion.[221]

The MMR case is especially hard and complicated when it confronts the citizen because of the following features:

1. It involves two distinct fields of expertise, medicine and epidemiology. The contribution of both, and their relative importance, have to be understood in order to navigate the domain.

2. Epidemiology is (or at least, was, before the coronavirus (COVID-19)-related publicity) a particularly esoteric domain, whereas most of us are continually in contact with medicine both through personal contacts and the mass media, but this does not reflect the relative importance of the two domains in respect of decision-making.

3. The origin of the anti-MMR movement arose out of reports of medical research, but the research does not seem to reflect the model of scientific research developed and championed in this book – craftwork with integrity – but publicity seeking, and, as it would turn out, a lack of integrity in terms of undeclared financial interests and medical procedures (though this was not widely understood at the beginning of the controversy, and we will minimise its role in our analysis).

4. Reinforcing the difficulty is that the medical journal *The Lancet* published an article bearing on the original claim, lending it scientific legitimacy. The article was later retracted. Given the paucity of the evidence reported

---

221 For an analysis in these terms and for an exploration of two sides of a vaccination debate, see Collins and Pinch (2005), especially Chapter 8. And note the way internet disinformation is a feature in the case of potential Covid vaccines (https://www.theguardian.com/commentisfree/2020/nov/13/daily-mail-anti-vaxxers-paper-covid-vaccine-mmr).

Figure 12.1: Front page of article starting on p. 637 of Lancet Vol 351 • 28 February 1998 as it now appears online as a PDF.

in the article, *The Lancet* itself seems to have been acting more like a journalistic outlet searching for readers rather than a medical journal supporting scientific values. This made it harder for outsiders to navigate the relevant body of expertises.

5. Many ordinary people had strong grounds for believing that they had direct access to evidence bearing on the danger of MMR. This evidence appeared to be of a quality equal or superior to the medical evidence. The emotionally persuasive evidence was the development of symptoms of autism in their own children soon after an MMR vaccination.[222]
6. The MMR revolt was bolstered by local parent-groups, which are themselves small groups.
7. Certain celebrities also had such personal experiences and spoke out about them, encouraging the vaccine revolt.

---

[222] The case arose at the height of the movement among certain social scientists to 'democratise' science and this encouraged them to side with these parents and, perhaps, encourage the parents to side with the revolt (see note 13). We will come back to this.

8. Perhaps most important of all, journalists, seemingly driven by the journalistic norm of 'balancing the story', gave equal weight to these personal experiences and celebrity statements on the one hand, and to the epidemiological evidence plus countervailing-medical arguments on the other.[223] Many journalists did not seem to understand, or to be willing to explain to their readers, even the minimal level of general science methodology needed to cast doubt on the initial medical claims – namely the tiny sample size, the fact that even this negligible evidence bore on the measles virus in the gut rather than MMR vaccine specifically, so applied equally to the single-shot measles vaccine, and the lack of any medical evidence for the relationship between MMR per se and autism – it was all a small number of coincidences mostly reported by parents and bound to happen where large numbers of vaccinations were occurring – while the single-shot measles vaccination continued to be recommended. Here are some quotations from the original paper:[224]

Onset of behavioural symptoms was associated by the parents with measles, mumps and rubella vaccination in eight of the 12 children. (p. 637)

Intestinal and behavioural pathologies may have occurred together by chance, reflecting a selection bias in a self-referred group; however, the uniformity of the intestinal pathological changes and the fact that previous studies have found intestinal dysfunction in children with autistic-spectrum disorders, suggests that the connection is real and reflects a unique disease process. (p. 639)

We did not prove an association between measles, mumps and rubella vaccine and the syndrome described. Virological studies are underway that may help resolve this issue. (p. 641)

If there is a causal link between measles, mumps and rubella vaccine and this syndrome, a rising incidence might be anticipated after the introduction of this vaccine in the UK in 1998. Published evidence is inadequate to show whether there is a change in incidence or a link with measles, mumps and rubella vaccine. (p. 641)

---

[223] See, e.g. Hargreaves et al. (2003).

[224] The paper is Wakefield et al. (1998). The tendentious remarks in the paper that seem to indicate a link between MMR and autism, it would turn out later, were probably driven by financial rather than scientific interests, but since this came out only later we do not make much of it here as we are trying to find a resolution that would apply at the time of origin of the anti-MMR campaign.

Not much in the way of scientific understanding is needed to comprehend the meaning and implications of these quotations; they may not have been comprehensible to everyone but should have been comprehensible to journalists taking the responsibility of conveying the significance of the scientific work to the citizen.

9. The MMR revolt took place in the context of a much larger and longer-standing anti-vaccination campaign promulgated on the internet. It was later subject to an organised disinformation campaign from a remote origin which bolstered it further as has been described in Chapter 9 with the illusion of intimacy likely playing a role.[225]

What would analysts and citizens need to navigate the world of MMR?

We must start from somewhere, and the authors of this book start from the firm belief that MMR vaccine did not cause autism in children and that the claim that it did was baseless. The authors of this book who are old enough to have been thinking about it at the time, can say that they believed this from shortly after the outset of the controversy, long before it was known that Andrew Wakefield, the medical doctor who started and promulgated the 'MMR causes autism' claim, had a financial interest in single-shot vaccines, long before it was widely known that his evidence gathering, such as it was, was suspect, and long before *The Lancet* paper was retracted. The grounds for our view were the paucity of the medical evidence and the misrepresentation of the significance of the 'evidence' of parents whose children first showed autistic symptoms after MMR vaccination. We think, then, that MMR is safe and that widespread vaccination with MMR is necessary for the continued prevention of the dangerous measles. Now we ask what analysts and citizens would have needed to make better decisions and undertake more sensible actions in respect of MMR or, at

---

[225] The Russian MMR internet campaign has been described above (see Chapter 9, An Example of the Malign Use of the Internet: Russian Disinformation Techniques). For the influence of the internet on anti-vaccination campaigns in general see, for example, Kata (2012) which can also be found at https://reader.elsevier.com/reader/sd/pii/S0264410X11019086?token=6 63122094C85C4D0C3575E391C52DE34DA4874ED053CE116FCA1FD90 9709F74A5E35D1BB0730E4B8B0A096654B763F29.

For a history of the growth of suspicion of vaccination in general see Blume (2017). The vaccination rate for MMR in the UK was around 90% before the Wakefield incident so we can feel reassured about treating it as a matter of a specific vaccine without worrying too much about the general context of vaccine distrust.

least, understand what better decisions and better actions would be? We have to phrase it this way because certain citizens, with a propensity for risk-taking and little propensity for public duty, even understanding the invisibility of the risk of MMR-caused autism, and even understanding the visible risk to their own children from measles should herd immunity to the disease be lost, might still choose not to vaccinate on the assumption that their children would be protected through the herd immunity engendered by large numbers of other citizens being ready to vaccinate.

Culture is the key

To anticipate, we are going to conclude that the MMR problem is an example of the grand problem addressed throughout the book: the problem of culture. We are going to conclude that the power, including power over the acceptance of vaccination, lies in the hands of those who can affect the national culture. We will conclude by suggesting how benign governments should act so as to shift culture in a way that supports citizens' understanding of SCD. In the meantime, we will look at some specific possibilities in respect of MMR in particular.

Restricting the internet as a solution to the MMR revolt?

While some of the impetus for the MMR revolt arose from the internet, restricting its content in respect of MMR does not seem a possible solution. This is because the possible link between MMR and autism was already being discussed openly in other mass media, including newspaper and television, not to mention *The Lancet*, and to suppress the discussion on the internet, however dangerous the extra energy given to it by the illusion of intimacy, would seem impossibly censorious. Remember that, at the time, Conservative politicians in the UK were ready to say that parents should be given freedom of choice of MMR or single-shot vaccine irrespective of the scientific evidence: to suppress such views would be politically impossible. There is still a role for regulation here, such as limiting the spread of posts about fake or dangerous remedies or vaccine substitutes, though even here the boundaries are hard to define.

Elementary science education and MMR

The early tradition in the 'public understanding of science' turned on the so-called 'deficit model', which held that citizens had a deficit of scientific under-standing that, if remedied, would transform them into science lovers. The idea of the deficit model became so discredited among social scientists that the term itself became an accusation that could close down an argument, and close it down in a way that, within the democratization of science movement, turned on the idea that citizens did not have a deficit of scientific understanding in the first place. But, of course, nearly everyone, including scientists, has a deficit of scientific understanding; the problem with the deficit model is that (a) the deficit cannot be remedied where science is complex and esoteric as it is in frontier science disputes, and (b) remedying deficits in scientific knowledge solves nothing where scientists disagree.

But scientific understanding has to be thought of as divided into a number of categories, many of which can be found listed and explained in the 'Periodic Table of Expertises'.[226] An important division is found on the third line of that table, between expertises that require a foundation of specialist tacit knowledge (interactional expertise and contributory expertise) and those that rest on a foundation of ubiquitous tacit knowledge (beer mat knowledge, popular understanding and primary source knowledge). Much of our criticisms of the democratization of science movement rests on their confusing the first general type with the second general type; it is taken that because citizens can understand a certain amount of established elementary science, they can also understand science that depends on specialist tacit knowledge which, in turn, depends on immersion into the small groups (core sets) that create and support it. Much of the technical disagreement that takes place in the public domain is frontier science and it is of this more complex type: it is too complex and too disputatious for it to diffuse among a population or to settle anything even if it did. In contrast there is an 'established elementary science', widespread deficits of which can be remedied to some extent, and to good effect. A smattering of established elementary science (EES) is the kind of thing that improves home maintenance and heating efficiency, an understanding of how temperature relates to tyre-pressures, fuel efficiency, and many other aspects of car-driving, and why vaccination works, the meaning of herd immunity and many other aspects of health and medicine. It would fit into the Periodic Table as a sub-category of popular understanding. The first elements of this kind of EES will be delivered in early socialization, as part of the culture which is more-or-less ubiquitous at the national level, and later, more formally in school. In later life a kind of continuing elementary education continues via the mass media and the internet and social media, though some of it is misleading.

In the case of MMR, there was a failure to make available the relevant EES even if people already understood the elements of the meaning of vaccination and how it works. This was because the EES was about a particular vaccine rather than vaccination in general – many people knew, or were soon informed by anti-vaccination campaigners, that vaccination could carry a slight risk, but the significance of the particular risk newly associated with the particular MMR vaccine was not something that would have been covered in school-level

---

[226] Collins and Evans (2007), Ch. 1.

education because it was not considered until Andrew Wakefield's claims came to light. Therefore, all the continuing elementary science education/socialization needed by citizens was going to have to be delivered by the mass media and, later, the internet. We have seen that the internet was later infiltrated by Russian disinformation amplifying the supposed 'controversy', we know that the information it delivered was not reliable, and the illusion of intimacy was amplified by the input of celebrities. We also know that the Wakefield campaign for a single vaccine fell naturally in line with the rhetoric of free choice and PAD, as do all anti-vaccine campaigns. In this case we had the perfect storm of Wakefield's attack on MMR, his financial interest in single-shot vaccines (as we found out later), and the supposed 'freedom to choose' offered by single-shot vaccines which fell short of full vaccine rejection; the argument that single shots would likely fall much further short of herd coverage than MMR was a subtle one turning, as it did, on predictions of human behaviour. Here certain prominent social scientists who sided with parents who were rejecting the vaccine can act as a litmus test for how the arguments went: they argued for the democratization of science and therefore the validity – in the sense that they should be taken seriously as scientific evidence – of parents' emotionally persuasive observational evidence when children did show autistic symptoms post MMR jab; they argued, perfectly correctly, since all vaccines carry a slight risk, for the possibility that there was some small risk in the MMR vaccine in spite of the epidemiological evidence that national autism rates had not increased where countries had introduced MMR vaccination; and they ignored the fact that the same sort of risk could be associated with the introduction of any medical treatment, or even new foodstuff, scare stories of this kind being constantly in the news. Science is not the exact thing portrayed in science myth – there is always enough flexibility to support a scientific 'fringe' and as many medical scare stories as you like. The social scientists should have understood this whereas instead, surprising given their academic backgrounds, they based their criticisms on the kind of perfect model of science that preceded the revolution in science understanding of the 1970s; they should have been aware that, to support the actual existence of possible minimal dangers, rather than their logical possibility, requires medical evidence, and in the case of MMR and autism, the trumpeted medical evidence was *not* sound – not sound to the point of scientific misconduct.[227]

---

[227] More concrete intimations of this were to emerge later. It ought to be a cause of shame to the profession that these prominent social scientists have never withdrawn or even been ready to reconsider their claims, presumably being supporters of Wakefield in his new and successful career in the United States among the conspiratorially inclined. [Even *The Daily Mail*, a right-wing nationalist newspaper, has retracted and apologised for its anti-vax stand and support of Wakefield's position (https://www.theguardian.com/commentisfree/2020/nov/13/daily-mail-anti-vaxxers-paper-covid-vaccine-mmr)].

But before the prominent social scientists, along with large parts of the rest of the population, could acquire their misplaced views in respect of the dangers of MMR, they had to get them from somewhere and, none of them being drawn from core-set of relevant epidemiologists, they (just like us) seem to have acquired them in the first place from that most important source of adult elementary science education, the mass media (possibly bolstered by the internet). And the mass media got things scandalously wrong in this case.[228]

The mass media, other analysts and MMR

The scientific understanding needed by outside analysts in this case is not something taught in (high) school, but it is the something more needed by those who are to present science to the public; it is a subset of primary source knowledge, the next category up from public understanding in the Periodic Table. This kind of knowledge is certainly needed by those journalists who take seriously their role as a check and balance within pluralist democracy, and it is needed, similarly, by social scientists of science. It is a basic understanding of what science is as an institution – part of which understanding has been developed in this book – and an elementary understanding of science's methods.

Analysts must know enough elementary science to be able to read and understand the significance of a paper like the one published by Wakefield et al. in 1998. In this case there is no need to understand the complex jargon of the field of virology and pathologies of the gut. It is necessary only to understand the significance of the quotations taken from the paper and set out above in paragraph 8 of the description of the special features of the MMR case. A social analyst who feels they are in a position even to start to comment on science, and a journalist who feels they can comment on such a case, must be able to understand that the numbers and claims in the paper do not warrant a serious concern about the safety of MMR vaccine and will not warrant it until much more evidence has been collected. They should understand that some coincidences between injection with MMR vaccine and the onset of autism are a statistical inevitability and carry no more information about a causal relationship than would such an inevitable coincidence between, say, first ingestion of a kiwi fruit and the onset of autism. They should be able to work out for themselves that such reports by parents, traumatic though the events may have been, are of no medical significance until there is evidence that their cause is MMR specifically (rather than just gut disturbance, which would implicate all measles vaccinations, not just MMR), and/or this relationship has been supported by evidence of epidemiological changes. Here, as is now well documented and certainly acknowledged by some journalists, the mass media got it wrong and were

---

[228] The prominent social scientists did not seem to rapidly move on to consider the evidence in more technical but nevertheless readily accessible, detail.

the amplifiers of the scare, led by the ethos of the so-called 'balanced story'; in this case, what they 'balanced' was parents' impressions with epidemiological evidence, thus putting a heavy anti-scientific finger on the scale and shifting the 'centre-ground' to the position that was also preferred, subsequently, by Russian and other sources of disinformation and misinformation. More elementary science education for journalists and others who take it upon themselves to comment on scientific disputes, and a greater readiness among such groups to take responsibility for the elementary science on which they comment, would be another change in the right direction.[229]

Non-specialists and MMR

But in the absence of a lead from journalists and the like, it is too much to ask ordinary citizens to work these things out for themselves: the nature of scientific experimentation and its statistical support is not taught in ordinary schools and is sufficiently subtle to have failed to be understood by a large subset of the professionals who should have been understanding it: including the journalists and those of the social scientists who got it wrong. To understand even at this basic level, one needs enough familiarity with science to have been able to access the original *Lancet* paper, read it, 'get the drift' of its argument and understand that the weakness of its statistical support disqualifies it as a significant contribution: this is impossible for the population at large for logistic if not for other reasons, even though it should lie within the envelope of scientific understanding – the 'primary source knowledge' relevant to a specific case – available to professional commentators who are, nevertheless, not medical specialists.[230] In a case like this, however, 'the public', other than the small number who might be exposed to training in a special case like a citizen's jury, have resource only to the expertise located on the fourth line of the Periodic Table – meta-expertise.

Meta-expertise has already been encountered in this book in the form of Figure 1.2: The fractal model of society, which is utilised again in Figure 11.1). The large oval at the top of the figure shows citizens' ubiquitous expertises; included among them, and emphasised by a box, is meta-expertise – the ability, where some important technical issue is involved, to choose how to rank the experts in the lower and smaller ovals; for example, the citizen must know that it is best to go the hospital when the body has a problem but best to go the garage when the car has a problem. Something has gone wrong in the world when citizens revolt against MMR vaccination because, instead of their consciously or unconsciously exercised meta-expertise guiding them to give

---

[229]   Boyce (2006, 2007) finds that the MMR story was mostly covered by journalistic generalists rather than science specialists.

[230]   It does not approach the level of 'interactional expertise' though one would hope that some specialist journalists would aspire to the acquisition of interactional expertise in some narrow areas of science.

special weight to the opinions of mainstream scientists and doctors when seeking an answer to the question of whether to vaccinate, they give more weight to the experience of parents whose children showed symptoms of autism after vaccination, or to celebrities who weigh in against vaccination, or to politicians who find the anti-vaccination stance to be a potential vote-winner, or to social media of unknowable provenance, or to wealth-seeking doctors or other scientists who are excluded from the mainstream core sets.

As the disagreements among coronavirus (COVID-19) scientists illustrate yet again, for politicians and policymakers, leave alone ordinary citizens, working out the mainstream scientific consensus, if it exists, is very difficult. The problem has been hugely exacerbated with the internet, which includes foreign-sourced disinformation on such matters, not least from Russian sources, which often purport to come from scientific experts. Of course, governments have scientific advisors who are meant to be able to filter the competing material, but this is an ever more complex task and one that should be done properly and accountably if it is convince a confused public. Collins and Evans in their 2017 book suggest the setting up of a purpose-designed institution as an organ of government, made up of scientists and social scientists – 'The Owls' – for reporting on the current substance and strength (represented by a 'consensus grade', something like A–E), of consensus in the scientific community in respect of relevant issues in the public domain; the substance and strength of consensus in the scientific community is a sociological fact not a natural science fact, though an understanding of the natural science debate would have to form the background to the inquiry. Such a committee would help to inform politicians (holding them to account for their claims about scientific consensus), citizens, journalists, and analysts of every kind. It should be borne in mind that the strength of consensus can sometimes be so weak that opinions coming from scientific advisory committees can be little more than that – 'opinions' – which would be given the lowest consensus grade by The Owls. Nevertheless, the argument remains that these opinions are, morally, the best that can be obtained because the aspiration that informs them is political value-free truth, even when it cannot be achieved.

Inevitably, the argument circles back to the internet. Consider the fractal model of society (Figure 1.2) once more. What we are arguing here is that citizen culture, perhaps supported by a body like The Owls, must come to *rely on* the right sub-groups from lower down the fractal. In this case we are referring to the sub-groups representing scientific expertise and the values that go with it; they must come to be the taken-for-granted first sources of understanding when it comes to day-to-day technological decision-making, such as that involved in vaccination choice. But why should this continue to be the case if the internet invades the relatively well-ordered world of constraining institutions represented by the fractal model?

With the internet, Figure 1.2 is no longer a faithful representation of society and SCD is no longer a viable political model. This is because the sharp boundaries to the sub-societies represented in the Figure are dissolving; the whole figure threatens to turn into an amorphous mass as knowledge liquifies.[231] The boundaries are now crossed by multiple links, 'weak links' in the sense that they can come from far beyond the mini-communities themselves but, nevertheless, potent links because the illusion of intimacy makes them just as persuasive as those that come from trusted sources inside the boundaries, and, from the inside, the way the boundaries are being transformed is not apparent. Society no longer looks like Figure 1.2 but is transformed, as shown in Figure 12.2. The dashed lines in panel B represent internet links, supported by the illusion of intimacy, coming from near and far, sometimes organised and sometimes not, and turning the well-bounded institutions and groups (such as panel A) that make up the fractal into fuzzy edged amorphous shapes reaching into mysterious social spaces (panel C); this is the social-structural counterpart of the liquification of knowledge. The well-ordered fractal model based on local trust (panel D), which forms the foundation of SCD, is replaced by something far more amorphous (panel E). With the internet, modern society, in spite of its size, once more has the connectivity of the Greek City State and allows PAD to function because seemingly individual opinions, or market choices, can be efficiently aggregated. But the dissolution of local boundaries means that the appearance of SCD would be no more than that: an illusion, because the structure has dissolved. Instead, as we know, the individual choice approach soon gives rise to control by aggressive centres of economic power and political power, as represented by the black stars.[232]

---

[231] The boundaries of most of the institutions were never quite as sharp as, for simplicity's sake, they are made out to be here; indeed, the revolution in our understanding of the nature of science associated with the 1970s showed that the boundaries of the institution of science were much more permeable to outside influences than had been thought. Nevertheless, the internet is a step change.

[232] In economics the dilemma was described by Karl Polanyi in his book published in 1944, who explained that, ironically, strong government was necessary to maintain free markets.

> the introduction of free markets, far from doing away with the need for control, regulation, and intervention, enormously increased their range. Administrators had to be constantly on the watch to ensure the free working of the system. Thus even those who wished most ardently to free the state from all unnecessary duties, and whose whole philosophy demanded the restriction of state activities, could not but entrust the self-same state with the new powers, organs, and instruments required for the establishment of laissez-faire. (p.147)

In politics, Robert Michels's (1911) 'Iron law of oligarchy' argues that democracy based on the idea of individual freedoms are always subverted by tendency to become controlled by oligarchical organizations. The causes

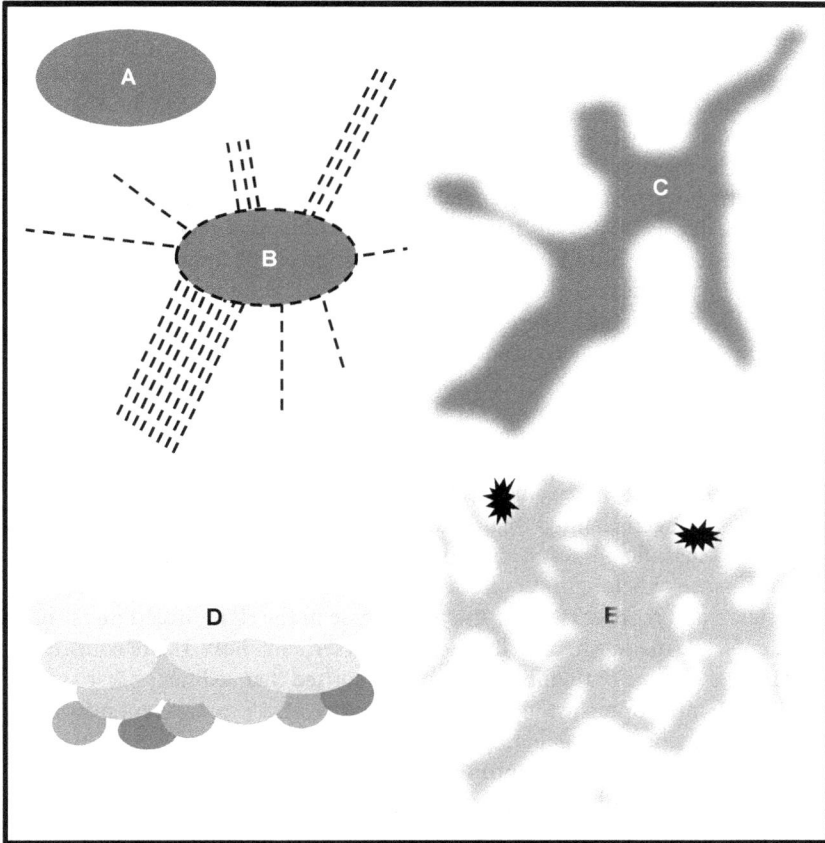

Figure 12.2: Core sets and other groups lose definition and structured choice democracy replaced by aggressive concentrations of economic and political power.

Unless these changes cease to happen, either through more careful control of the internet, or deeper understanding of how truth-making works, including the way it depends on face-to-face interaction to develop trust, the very idea of citizen meta-expertise will dissolve.

---

and mechanisms are very different to what happens now that powerful organizations can gain control over individuals' thinking via the internet but the parallels with Polanyi's analysis of economics are suggestive and equally suggestive of the need for new controls.

Developing meta-expertise in MMR

In Chapter 1, we introduced the concepts of 'trust' and 'reliance'; these concepts underpin the notion of meta-expertise in respect of this case and the argument of the book as a whole. Reliance, remember, is the unthinking, taken-for-granted kind of trust, which nearly all our actions depend on, which we mostly develop in the course of early socialization, and which is the foundation of our knowledge as citizens. Without citizens trusting, and relying on, the kind of institutions that promulgate sound knowledge about issues like the safety of MMR, just as they rely on the institutions that take them on taxi rides and that sell them foodstuffs, there seems little hope of solving such problems by extracting them from the internet liquidizer while still in one piece.

Democratization of science as a non-solution

Before looking at positive solutions to the problem, let us go back to democratization. It has been claimed that democratizing science will make citizens feel part of the scientific process rather than alienated from a hostile scientific elite. This new sense of belonging could raise the status of science once more in popular esteem and reorder citizens' meta-expertise in the right direction. Perhaps a case could be made where 'delineated cases' (see Appendix 3), are concerned, but MMR is not one of them.[233] We have established that ordinary citizens have nothing scientific or technical to contribute to the MMR case except misplaced and misleading personal experience. It might be that if citizens' understanding of the matter was welcomed into the fold, citizens would indeed feel warmer toward science, but that new valuation would have been bought at a heavy cost: accepting citizens' estimate of the value of medical interventions in place of estimates based on medical and epidemiological science.[234] In other words, the re-evaluation of science would be a sham – it would be self-defeating. Remember, that as argued in the previous chapters, there is no tension between democracy and certain elite institutions, since SCD depends on elite institutions as does the very idea of technical expertise.

---

[233] It can also be the case that citizens can offer certain kinds of data collection as in one interpretation of 'citizen science', or local knowledge of corruption (described as 'local discrimination' in the Periodic Table of Expertises). Encouraging citizens in all these kinds of cases could serve to give them more confidence in science but it must not be mistaken for other kinds of more esoteric scientific input.

[234] This is not to say that citizens' systematic anecdotal reports of, say, medical symptoms such as 'long Covid', should not be counted as a delineated case, at least triggering some attention from elites but not competing with elites in the matter of scientific evaluation.

How to change culture and who should be doing it?

So let us return to the general problem of changing culture in beneficial ways – that is ways that promote truth to the centre of decision-making. We already know something about how culture can be changed by those with sufficient resources and skills because we have discussed two important examples in this book. Both examples use the internet and control over mass media outlets where possible. The first of these is digitally enabled disinformation and other influence operations (including those of far-right domestic populations).[235] Foreign powers believe they can disrupt the political and medical culture of Western democracies by feeding disinformation into social media and the like, making use of anonymity and the illusion of intimacy which it facilitates. The easy anonymity and identity faking are new to electronic communication; old-fashioned spying required far greater resources and ingenuity and was far more open to exposure and danger.[236] We do not know how successful this attempt to influence culture has been, but we do know that the powers that use it put a lot of resources into it so they must think it is having some effect. The success of the second method, which uses mass media more heavily, is evidenced by the shift of the perceived centre-ground of American politics and the fact that actors like Andrew Breitbart have not been shy about explaining its centrality and the means by which it has been accomplished. Again, across a number of central and Eastern European states, there is a trend towards increasing concentration of media ownership, with oligarchs blatantly buying up media outlets and exerting editorial influence upon their output. Governments who want to shift culture in a benign direction or prevent it being shifted in a malign direction, must understand these methods and react to them.

Once upon a time there was at least a chance that analysing the problem and explaining its solution might have led to benign governments considering how to execute policies to shift cultures in the right direction. Unfortunately, writing in the last months of the year 2020, it is not at all clear that Western governments would execute such policies even if they came to their attention and they understood the consequences of following them or otherwise. One possible consequence of not following them, we have argued, is 'The West', instead

---

[235] Actually, the success of these operations might be better established than we are suggesting here – for example, see the analysis of the electoral manipulation work of the firm 'Cambridge Analytica' in recent documentaries such as Netflix's *The Great Hack* and UK Channel 4's interview with targeted generation of anti-Clinton opinion in the Republican's 'Project Alamo', mounted during the 2016 US presidential election campaign (https://www.youtube.com/watch?v=Jg9QaUyQ3lc).

[236] As Rid (2020) points out, current disinformation tactics have a long history in the form of active measures, but with the rise of new digital technologies they became *more active* and *less measured*.

of winning the Cold War, conceived of as a battle between totalitarianism and democracy, losing it, in a pathological surge of proselytization for economic free enterprise and the associated political freedoms, the false idols of negative freedom, which, in turn, are the ideological hunting grounds of the far right.[237] This surge leads to a concentration of power on the political and economic right, with private enterprise firms and their political puppets controlling political life, with decreasing limits on the power of government and economically driven pressure groups, supported by the rhetoric of populism and leading to ever greater economic inequality. In many countries, this would be the demise of the economic 'Post-War consensus' and the rise of a new kind of political and economic monoculture which circles back to a totalitarianism of the right. There can be little doubt (this is written eight days before the American election of 2020), that President Trump, supported by the Republican Party, prefers such an outcome to loss of power (in the Postscript we will reflect on the meaning of the election's outcome). Though things are less extreme in the UK, the continued attacks on the rule of law and the other checks and balances by the UK government point in the same direction.

If these governments are truly set on this path and if they are not called to account in the next elections, or by their own parties (very unlikely in both countries, especially in the USA), then we are merely spitting into the wind unless there are other forces with a more benign outlook and the wealth to do something about it. In the meantime, let us spit and hope the wind changes direction in time.

Civic education, democracy and science

Governments which want to preserve democracy should understand 'The Law of Conservation of Democracy', which, with a bow to physics' Law of Conservation of Energy, states that you cannot take more democracy out of a society than you put in. The claim is that if democracy is to be preserved then citizens must be taught to understand democracy so that, if they want to preserve it, they know how to vote, at elections, for parties that also want to preserve it. Obviously, the Republican Party and President Trump do not want to preserve it, but unless the US population understand what they are losing by re-electing him, democracy proper in the USA is unlikely to last. Of course, Republicans will claim that democracy is to be identified with maximum individual freedom (e.g. to carry guns etc), so will present a vote for guns and Trump as a vote for democracy not against it, but that as we have explained, is to misrepresent democracy or to encourage a form that is self-defeating.

---

[237] Europe should not be seen as independent of the USA in this regard: the density of the alliances, military, economic, political, and in terms of security, are too great.

Civic education

If we had a benign government, one way in which a proper understanding of democracy should be bolstered is through a revival of civic education that explains the positive role of public institutions and the necessary role of government in society. But wouldn't an educational system that was tasked with explaining the positive role of government in society be indoctrination into the ideology of the left – a flirtation with 'socialism'? Do citizens not have the right to choose freely what they mean by democracy?

The argument is a catch-22: if you already believe that democracy is defined by maximizing individual freedom at any societal cost, then the teaching of anything that does not support that view, instead of leaving citizens to choose what kind of democracy they want, is, by definition, a violation of democracy. Therefore, the argument for better civic education can be supported only if democracy really is something other than the nurturing of maximum individual freedom. But this takes us back to the charge of 'socialism' – and so on. We just cannot move forward without agreeing that modern society is, and should be, both constraining and enabling. As explained in Chapter 11, society keeps us safe (e.g. in most countries by radically limiting the sale of guns, and in nearly all by setting traffic regulations, controlling the sale of drugs, regulating the safety of garments and furniture), and society also provides the cultural infrastructure that makes us more than simply a kind of animal. The catch-22 is broken because the idea of unlimited freedom is untenable both philosophically and because it leads to a rightist economic monoculture with power being as concentrated as in any totalitarian state; democracy defined as maximum individual freedom leads to the demise of democracy – that is the catch-22 that we should be worrying about. That catch-22 could be leading to our self-inflicted loss in the Cold War as, for the sake of power and electoral victories, we sacrifice the very values for which the Cold War and previous hot wars were fought. And worse, as this book has argued, the speed at which it can happen has been hugely increased by the rise of the internet and the spin of the knowledge liquidiser. So more civic education is the right and urgent policy for governments who want to maintain democracy and are not afraid of being voted out of office by a newly and properly educated public.

The main content of that civic education would be the role of society in enabling our abilities, and the role of government in regulating societies in ways that make our lives efficient, safe and secure against too much concentration of power. A slightly more advanced form of such civic education would include an analysis of the two kinds of democracy outlined in Chapter 11.

Another part of that content would be an explanation of the nature of science and the role of science in society. It would lead toward a relative distrust of what is found on the internet and a growth of trust of the institutions of science. It would lead to a proper ordering, in the minds of citizens, of the elements of the fractal in Figure 1.2; this is nothing more complicated than knowing that to fix one's car one should take it to a garage, not ask a celebrity.

Explaining science to citizens

Explaining science to citizens is not an easy thing to do because of the distorted image that is already traditionally entrenched in society. This image is based on the famous discoveries of science – the crown jewels, often triumphant analyses of the movement of a few objects in nearly empty space, such as the movement of the planets or the interaction of sub-atomic particles. The image is science as a producer of certainty, or even a producer of a kind of superior magic and mystery. But very little science is like this, and science in the public domain is generally far from it because of the complexity of the matters it deals with. Mostly, science is craftwork with integrity. Coronavirus (COVID-19) science is a perfect example – even the better part of a year into the pandemic we are just discovering the extent to which aerosol transmission is important; we do not know why the outbreak spreads in the way it does nor how to control it except by instituting draconian separation between people; and, above all, outside of authoritarian regimes such as China, we do not know the extent and the duration of the population's readiness to follow government guidelines and rules nor how that readiness varies among different social groups. But in spite of all that ineffectiveness and uncertainty, citizens are realizing that they desperately need the science to inform political decisions, even if we know that things will likely go wrong. We know that when committees of scientists offer advice to governments, they are offering the outcome of the best analysis they can do, maximally unaffected by political or financial interests. That is why the MMR case was so demanding – those last provisions did not apply to those who started and promulgated that scare.[238]

Craftwork with *integrity*, not magic or certainty, is what we should be demanding from our science. But science should be our icon: craftwork with integrity should go into every process of decision-making, not just scientific decision-making. Craftwork with integrity is what we should be demanding from our politicians; politicians' skills and abilities must comprise more than the ability to win votes. The ability to win votes is one thing, the ability to make decisions that will work for the good of the country is another; recent events in the US and UK have shown how different they are today, and it is the internet that is separating one from the other by an ever greater distance.

---

[238] For an analysis of the response to coronavirus (COVID-19) in the UK that documents both the uncertainty of the early months of the pandemic and the problems that arise when policy-makers expect certainty from science, see Evans (2021).

Therefore, the other thing that could be part of civic education is the kind of understanding of the internet that has been developed in this book. Citizens need to understand that face-to-face communication is the basis of reliable *non-transactional* trust where *dangerous and difficult truths are at stake*, and that institutions that recognize this model are worthy of special respect. We should learn that the illusion of intimacy generated by the internet is just that – an illusion. We, or whoever we learn from, has to know about the difference between kinds of groups and the importance of core-set values in the groups we should be relying on and trusting.

Changing culture outside the classroom

The creation of culture starts in the home, moves to the school, but continues through adult life, where it is energized by the mass media and the internet, and maintained and reinforced through the actions and spoken utterances of every citizen. Everything we do is a contribution to what we, as a society, believe and rely on in a taken-for-granted way. We have lost control of the driver of cultural control and change that is the internet. Before the internet, cultural input was much more a national matter than it is now. Admittedly the film and television industry were cross-cultural influences, with American influence strong in English-speaking countries, boosted by consumer-culture infiltration known as McDonaldization or coca-colonization, but that was slow and visible. Nowadays the internet is a global spider's web with anonymization rendering its reach invisible without focussed investigations. To this extent, national culture and the influencing of national choice is no longer solely a national matter.

This change suits those Western politicians inclined to populism because the internet can be used by foreign powers to encourage the erosion of Western democracy and thus strengthen their cause. Therefore, at the time of writing, we see Western leaders blocking or discouraging investigations of Russian interference in elections. Once more, so long as the leadership of the Western nations prefer power to democracy, and so long as they are ready to endorse long-term defeat in the Cold War, our pointing this out is spitting into the wind.

But if a leadership was elected that wanted to reverse the trend, what could it do in addition to constraining the internet (while evading the charge of censorship), and encouraging civic education in the classroom? It would, of course, encourage those inquiries into outside interference in elections rather than discourage them. And then it would take its culture-changing lessons from people like Andrew Breitbart.

Andrew Breitbart was the founder of 'Breitbart News', an influential disseminator of right-wing disinformation and 'fake news' in the US. His analysis of the key ideas that move and mould us collectively was something understood by dictators in mass societies long before the internet, with Hitler and Goebbels being the icons, but the point being equally recognised in communist societies:

ownership of the means of cultural production affords a large measure of control on the ordering of society.[239] Breitbart seemingly appropriated aspects of Antonio Gramsci's theory of hegemony, and the sociology of knowledge and culture in general, and re-tooled them for a contemporary right-wing, social media-saturated, audience.[240] Breitbart's most famous axiom was the idea that 'politics is downstream of culture' by which he meant that controlling and owning the means of cultural production was more influential than the formal institutional power of politics, in terms of shaping and steering the ideas and issues that most people think about, most of the time. In effect, he anticipated that politics would be increasingly susceptible to the frames and techniques of persuasion familiar in entertainment media. He elaborated his thinking on this point in his memoir thus: 'The left wins because it controls the narrative. The narrative is controlled by the media. The left is the media. Narrative is everything.' This analysis has become an important touchstone for those that have sought to harness alternative media sources in order to try and undermine what they see as the hegemonic power of the MSM, with the intent that it should provide a pathway for political impact.[241]

We do not agree that the left controls the media or the narrative – far from it (unless you want to redefine the 'the centre' as what is traditionally thought of as the far right) – and we do not mean to say that institutions like the BBC have the kind of control that the word 'hegemony' often connotes, but we do mean that we see culture as forming society and, therefore, upstream of politics, and we do agree that the mass media can be a powerful contributor to culture.[242]

As mentioned above, in a number of central and Eastern European states there is increasing concentration of media ownership, with oligarchs buying up media outlets so as to exert editorial influence on their output. Simultaneously, in the UK, a right-wing government is attacking the finances of the BBC.[243] In America, of course, the Trump administration continually attacked the media

---

[239] The pre-eminence of culture is, of course, the very substance of the sociology of knowledge, the root perspective of several of the authors of this book.

[240] For Gramsci's theory of hegemony, see, for example, Bates (1975).

[241] See Breitbart (2011, p. 4).

[242] In so far as the media has hegemonic power in Western democracies, it seems far more likely to be right-wing messages than left-wing messages that dominate; see for example the distribution of political sentiment in British newspapers.

[243] Terrifyingly, just a couple of days before the 2019 general election, right-wing newspapers in the UK were lauding the fact that Boris Johnson was proposing to reconsider the justice of the licence fee that supports the BBC; this was seen as a positive move not a potential disaster for pluralist democracy; it indicates a huge lack of understanding of the nature of democracy by a large proportion of the population.

that was not controlled by his regime or its allies, as presenters of 'fake news'. Once more, in the West we are, so it seems, busy trying to lose the Cold War by subverting those very institutions that maintain the integrity that is at democracy's heart. Thus, the first concrete thing that the US and the UK could do if they want to preserve their democratic culture is move away from the populist instinct to attack institutions that constrain power; instead we should be reinforcing those institutions and strengthening their role. This is not as crude as buying up media outlets to control their output, it is merely maintaining the existing media institutions that have the notion of truth at their heart, rather than displacing them with puppets of right-wing regimes. Among other things, it is the survival of such institutions that most recent wars have been fought for. Public service broadcasting is, perhaps, the most important institution of this kind.[244]

Given the influence of wealth on the direction of mass media outlets that nowadays promote right-wing politics in the successful attempt to shift the centre-ground of politics to the right, there might even be space for private money coming from more centre-inclined sources to finance Fox-style media outlets but with a traditional centre-ground agenda. Whether it is public service broadcasting or new privately financed media outlets, the idea is to feed into adult socialization what it means to live in a democracy: to enable ordinary citizens to learn to rely on what we have to rely on if we are to maintain democracy through future elections.[245] To reinvent the story of America in Hollywood and other sources of public discourse in a way which sees it less as a frontier nation borne of rugged individuals, competing all against all, and more as a triumph of social cooperation and organization. The *American Soldier*, it turns out, is, above all, a group member;[246] the Manhattan project was an object lesson in organization; American icon, Henry Ford, was an organiser; as the frontier closed, the farmers beat the cowboys and it was this that made America great. Fences, fertilizers, schools, society and DARPA-like government agencies, not rugged individuals; that is the story.

---

[244] 10 November 2020: Oliver Dowden, UK Conservative Party Secretary of State for Digital, Culture, Media and Sport, sets up an inquiry into public service broadcasting, asking if it is necessary in the digital age. It is, of course, more necessary than ever, but watch this space with trepidation.

[245] See this website for the potential as realized by George Soros: https://www.opensocietyfoundations.org/what-we-do. Note, also, that as described in note 220, private organizations such as Facebook have recently been moderating President Trump's stream of false social media claims with vetos or warning signs. These examples show that it is at least possible for business to act as upholders of genuinely centrist political culture.

[246] Stouffer (1949).

The substance of democracy

The trouble with the terms 'right', 'centre' and 'left' when it comes to politics is that, formally, they are relative terms. This makes it easy for them to be continually redefined and used as argumentative resources in political debate with meanings being shifted as convenient.[247] As discussed, Republican activists have been engaged in a successful campaign to relocate the 'centre' of American politics, as perceived by many of the electorate, in a rightward direction. This leaves the traditional centre vulnerable to attack as a 'left' position. Since his election, President Trump has been heard proclaiming that the USA would never become a 'socialist nation'. But what do he and his cohorts mean by 'socialist'? We may be sure it is not what most socialists mean by socialist; for Trump supporters, and those they can fool, 'socialism' is anything that can be portrayed as limiting individual freedoms.[248]

Here we are going to argue that the centre-ground of Western politics can and should be defined, once and for all, in an absolute way, as SCD, just as 'far left' and 'far right' have widely accepted meanings. 'Far left' generally means communism, with socialism, for communists, being merely a way-station. Communism and SCD share the view that citizens are products of their society and, in a strange and distorted way, they even share that position with the far-right view that 'the folk' are to be identified with their race and nation; unconstrained negative freedom is a different form of far right, more typical of America. The difference between the far right and far left versions of citizens being products of their society, and SCD's version is that in the former the influence of the society is to perfect citizens and make them uniform, with deviance being treachery, whereas SCD celebrates the different things that citizens become after secondary socialization, as illustrated in the fractal model of society, and how these differences contribute to the maintenance of an equitable society. Things get complicated because the ideal of unrestricted negative freedom and the idea of the sovereignty of individual rational choice leads to the idea of 'the will of the people' and a leader who can represent it and define resistance to it as treasonous.

---

[247] As 'persuasive definitions' (Stevenson 1938).

[248] Another meaning of socialist was exploited by Trump in his appeal to Latino voters. Many of them were refugees from Marxist regimes in Cuba and Venezuela and, with Trump claiming that Biden was under the thrall of self-declared 'socialist' Bernie Sanders (whose policies are actually a little to the right of the post-war welfare and economic settlement in the 1950s UK), and their still having social networks stretching via the internet back to those countries, 'socialist' for them meant all that was bad in the regimes they had escaped from.

Now let us suggest how SCD ceased to be the obvious and widely accepted meaning of 'centre' in Western politics. The argument begins with a speculative historical thesis concerning 'The West': the rightward shift of what is perceived by many as the 'the centre' has been fuelled by the conflict with communism. This is now merely a story for those younger than about forty and for whom, say, the fall of the Berlin Wall, in 1991, was not an astonishing lived event of huge political significance. What we are suggesting, is that episodes and events such as the nuclear arms race and the spying that triggered it, the creation by military force of an 'Iron Curtain' between European countries under Soviet thrall and Western Europe, McCarthyite anti-communist paranoia in the USA, the Cuban missile crisis, and the building of the physical Berlin wall in the early 1960s, events still vivid for the older generation, have triggered, among certain powerful political actors and their supporters, a pathological reaction to anything that involves state sponsored cooperation, or even the recognition of the value of society in the creation of individuals. This reaction rebounds all the way to the idea that any constraint on freedom of individual action is bad, aligning with Rousseau's version of politics and, of course, PAD. Theoretically, the idea was bolstered by free-market economics, championed by Reagan and Thatcher (very much members of that older generation), utilized by Reagan to destroy the Soviet state by confronting it with an economically unwinnable arms race, and symbolized by Thatcher's notorious 'There is no such thing as society'.[249] It might seem that if communism was as bad as it appeared to be from the late '40s to the 1990s, then the further from it one could get the better. Of course, businesses with an interest in minimizing regulation on their activities also have an interest in keeping this sentiment alive, whether or not the historical speculation is valid.

But, as pointed out, the opposite of communism is not maximum freedom because maximum freedom is self-destructive: at best it leaves citizens unprotected from danger and at worst it circles back in various possible ways to dictatorships of the far right. The defence against the far left is not the far right, not maximum negative freedom, but is the centre, and that centre is SCD. But that realization involves recognizing and celebrating the way that individuals are made by their societies, and being proud of their collectively honed abilities, even if it is not in the uniform way celebrated by communism.

Ironically, it is PAD that might, at first sight, appear naturally aligned with the left because of its proclaimed emphasis on 'the rights of the people', but unfortunately, in most communist regimes as they are instantiated, the people's desires are transmuted into dictatorship, and the peoples' rights to plural opinions

---

[249] '[T]here's no such thing as society. There are individual men and women and there are families. And no government can do anything except through people, and people must look after themselves first.' (*Woman's Own*, 1987 Sept 23rd)

curtailed; the same happens in the rightward direction when PAD aligns with populism. As we have also seen, PAD simply does not work when it comes to technical decision-making: the interpretation of the MMR revolt by certain leading social scientists, who, we may be sure, are of a politically centrist or even left-wing persuasion, demonstrates this. Furthermore, PAD simply does not work when translated into the maximization of negative freedom: freedom of choice without interference from the state or a concern for the common good is, in an absolute sense, disastrous, morally, philosophically, politically and financially.

What we mean by 'the centre' is, to repeat, congruent with SCD, a democracy that allows consideration to non-majority opinions and checks and balances on power. What we mean by the centre of politics is what, in 'the West', the Second World War and the Cold War were fought for. Under this meaning of centre, political debate between centre left and centre right turns on the choice between a range of policies that share the ideology that government, state, taxes, public education and society are necessary things. On the *left of centre* these policies would include public service broadcasting; affordable universal health care and accessible high-quality education for all; a determination to prevent the acceleration of economic and other kinds of inequality in society; continual consideration of the level of taxes with tax-rises, especially progressive taxes, an economic possibility; a politically independent judiciary; and some nationalization of services such as trains and the postal service on the grounds that long-term investment in efficient nationalization of some services is necessary for business development, equality of opportunity even where populations are sparse, as well as the reduction in the environmental cost of other forms of travel.[250] On the *right of centre*, some or all of these would be rejected, but not by claiming them to be 'socialism', sympathetic to communism, and in conflict with basic freedoms; they would be just a different set of economic and environmental choices within the envelope of a politics that recognized the vital role of society and government under SCD.

All would agree that there should be a science driven not by primarily wealth creation, or the political attractiveness of its findings, but by craftwork with integrity independent of politics and financial interests; a science that could

---

[250] Perhaps the government-sponsored loans given to help businesses survive at the time of Covid should, instead, be grants purchasing shares in those businesses with the long-term aim being a profitable return to taxpayers, a scheme suggested in the Financial Times of 6 Nov 2020 (https://www.ft.com /content/b81f2bc8-bbe7-4c86-98c5-2b5f80ba70d9?accessToken=zwAA AXXIGLXAkdO4HyvIu-dMhtOYxStfgLpw2Q.MEQCIDhMk8ZVEqPg WOZY5yS8NMn6V9jhRuV3lsgl9wj1_Fm5AiBC6-K3hS2n9zKnqPZ3lhJ kzrWAuveq1jldaLCJxTI1lg&sharetype=gift?token=d37fb4a6-7d32-4475 -81f6-a5c21989d739). Again, such schemes, potential enhancements of the interrelationship between government and business, should be considered on economic grounds, not as an indication of creeping socialism.

give leadership to decision-making of all sorts; a respect for scientific expertise that would foster a nation with truth at its heart.

To repeat, this comes with the recognition that the idea of too much negative freedom – maximum freedom from regulation and control by the state – is both philosophically untenable and undesirable in practice. It is philosophically untenable because society provides our language and all the rest of the understanding of life that divides us from the animals, continuing in more formal education regulated by the state. It is undesirable in practice because we want our traffic regulated, our children to be protected from inflammable nightclothes, our shower glass not to be lethal, and the dangers of tobacco smoke to the individual, and global warming to all of us, to be explained and acted on irrespective of commercial interests. SCD, unlike PAD, recognizes that citizens are not free-floating choice-makers, but their opinions are formed by the groups in which they are embedded and, nowadays, by digital campaigns designed to create political influence. Furthermore, in spite of financial free-enterprise ideology, even economic freedom lovers want the monopoly power that it gives rise to, to be controlled. They even want a state that supports technical industries to the lavish extent that America supports its industries (ironically, under the banner of 'free enterprise'). All these things could be described as 'socialism' but, if these things are socialism. then nearly all of us, including the leaders of wealth-creating companies, are socialists. This much socialism is not the far left, it is the centre.

Conclusion

So now we can say what a Western government that does not want to bring about its final defeat in the Cold War would include in its programme of civic education and what it would expect of its well-supported public service broadcasters and its admired free press. It would expect all of them to support the checks and balances of a SCD including scientific expertise. It would expect them to encourage and explain a science that is founded on craftwork with integrity, not sure success. It would value science as an institution with truth at its heart, setting an example to a democracy with truth at its heart. It would expect public education to include an explanation of the essential role of society in making us human not animal, and the role of governments in regulating our lives for efficiency and safety. It would even take an active role in encouraging and supporting economic growth with regulations designed to prevent the pathologies and inevitable corruption of markets that are too free, as well as the short-sightedness of investment decisions which look no further than the short term. These understandings should come to be the basis of what the citizenry come to rely on and, if they do come to rely on them, they will elect democratic governments in line with The Law of Conservation of Democracy. Let us hope we are not spitting into the wind.

Once more: the argument in sum

Finally, let us, once more, draw together the themes of this complex funnel of an argument concerning science, communication, truth and democracy. We started with the idea that, other things being equal, truth is better than lies. Whenever we may seem to be taking a more unapologetic line about a political position or set of actions, less hedged about with qualifications than academics normally think proper, this is the premise on which we are building. For example, we suggest science should be a central institution in any society that favours truth over lies because science is the institution – the form of life – that, most clearly, is founded on the idea of truth not power: science sets out to be a truth creator – that is its purpose; whether it can achieve it or not is a separate issue. We show that face-to-face communication is central to the aspiration of truth creation in science, and we argue that face-to-face communication is a necessary if not sufficient condition for all 'difficult and dangerous' truth creation, including political policymaking, and that science should offer an object lesson in respect of all kinds of decision-making as well as being a check and balance on political power.

We show that the internet has shifted the balance of communication in society from face to face to remote and we show the dangers that this brings with it. The new kind of communication liquifies knowledge and makes the taken-for-granted knowledge of societies – what we *rely on* to live in society – vulnerable to control from the outside. We show that remote communication is especially powerful because it can disguise itself as the trusted local communication – 'the illusion of intimacy'.[251]

Science and democracy, we have argued, are intimately linked.[252] But this link depends on a proper understanding of democracy. Science and democracy are antithetical under a popular assertive model because the elitism of science renders it opaque to the citizenry; the only resolution under this model of democracy is magical thinking. But the popular assertive model is antithetical to all democratic checks and balances, including the rule of law, which depends on elite analysis. Indeed, the popular assertive model rests on a false model of humankind and is self-defeating.

The proper understanding of democracy is SCD, which depends on elite institutions just as much as it depends on the ultimate decision of the citizenry in respect of whether those elite institutions have done their job – no magical thinking is required; it is the only model of democracy that makes sense, the only one that is compatible with a truth-making science and the only one

---

[251] It could be said that we have finally learned that more information is not always better; it is the translation of information into knowledge that is always the key.

[252] This is an old argument going back to Robert Merton, but we have recast it, finding it needs, among other things, a proper analysis of democracy.

that can resist the siren call for maximizing negative freedom. Much of this is summed up in Figure 12.2.

Western democracies are obsessed with their conflict with communism and consequently obsessed with a notion of a false freedom and anything that can be labelled 'socialism'. The result in the US is that half the population cannot distinguish between the power to maintain a façade of sham freedom on the one hand, and democracy on the other. To rescue these democracies from themselves, we need a public education, which will explain that *SCD* is what democracy means and will explain the roles that science, and the model of communication used by science, play within it. We have tried to find ways of bringing a viable democracy back to the West and avoid final defeat in the Cold War. The best we can think of is a vigorous re-introduction of civic education in schools and a safeguarding of truth-driven public broadcasting, with, perhaps, private investors supporting centre-ground politics media to counteract the investors who are shifting the centre-ground to the right. What better use could there be for the huge fortunes acquired by those who have benefitted from the shift to remote commerce and communication consequent on the COVID-19 lockdown? The winners have largely been those who have also benefitted from state input to the technologies they use and are of a generation that understands the importance of scientific expertise: why not put money into preserving the culture that supported their enterprises rather than allowing those same enterprises to lead to its demise? This would be an inspiring example of the voluntary setting right of the initially unforeseen consequences of technological change. And no one is in a better position to act with the kind of urgency that is needed. The matter is urgent; we have been given a breathing space by the Covid pandemic, which has revealed to some the power of actions driven by a commitment to honesty: actions driven by a commitment to honesty save lives and save democracy. The breathing space is short: cultural change takes time; here it must happen quickly.

The November 2020 Election in the USA

The morning of Wednesday 4 November 2020 was the most politically shock-
ing few hours in the lives of many people, including the authors of this book.
After a couple of weeks with hope of a Democrat victory bolstered by the opinion polls, for most of that Wednesday, it looked as though Donald Trump had
won the election after all and was headed for a second term. For those people, it
looked like the end of Western democracy without any obvious way of regain-
ing it: a second term of Trump would signal the destruction of truth itself, not
to mention the institutions that turn on it. No potential change of this magni-
tude or evil had been seen in Western societies since the 1940s.

This Postscript began to be put together on that morning. In the USA, civil
unrest was a real possibility if things showed signs of turning round and, in
the longer term, if things did not turn, chaos in large parts of the rest of the
world seemed inevitable: our children, if we had them, would have to live in
a world informed by a very different set of expectations. The world we thought
we lived in would have turned out to be better described as a brief golden age.

It seemed that the Postscript might as well be written on 4 November because
nothing much in the message of this book would change depending on which
way the outcome would eventually fall: the fact that around 50% of the US
population could still vote for Donald Trump already illustrated the Law of
Conservation of Democracy – without a population that understands democracy, democracy is unsafe. The way the vote finally fell might have huge consequences for the value of this book – spitting into a hurricane or spitting with
the breeze – and enormous consequences for America – the end of democracy
or a chance to pull back from the brink – but the difference would be the flap
of a butterfly's wing – and flap of a butterfly's wing does not lend itself to sociological analysis even if the consequences do. By 'democracy', of course, we do

How to cite this book chapter:
Collins, H., et al. 2022. *The Face-to-Face Principle: Science, Trust, Democracy and the
Internet*. Pp. 219–222. Cardiff: Cardiff University Press. DOI: https://doi.org/10.18573/
book7.n. Licence: CC-BY-NC-ND 4.0

not mean something that takes the elected party to represent as 'the will of the people'.[253]

According to the opinion polls, Biden's Democrats had been strongly ahead of Trump's Republicans in the last weeks of the campaign, and this was a surprise to many Democrats; they sensed that Trump's charismatic approach was going to be hard to defeat in the context of the US population's understanding of democracy. Biden's lead in the polls could, however, be explained by Covid: Covid had made the electorate realize that they needed science more than they needed post-truth and alternative facts. The disease, it seemed, had choked the knowledge liquidizer with the dead![254] But the election night surveys of opinion seemed to show that things were spinning up again: electors considered that promises of a rosy economic future outweighed the potential demise of truth and that, as reported in media interviews in Florida, 'socialism' was the thing to be rejected at all costs (to be feared far more than the demise of science and truth). Nevertheless, by Thursday, Biden's chances had improved in anticipation of the counting of the unprecedentedly large number of postal ballots. It is just possible, then, that it was the Covid virus that caused the butterfly to flap its wing, but we will probably never know, and it is not important since things could easily have gone the other way, and, more terrifyingly, could go that way again in the future. And make no mistake: it is Western democracy that is at stake. Analysis of the record of the Republican Party over the last decades, in terms of a series of indicators of democratic actions, shows that, unlike the Democratic Party which has remained steady, it has increased its readiness to demonise opponents and accommodate violence, reaching toward the kind rhetoric more typical of populist states like Hungary. The increase in authoritarian actions has been particularly notable over the last decade and very steep since the start of the Trump era.[255] To feel what it means for it to have been a narrow difference in a chaotic system that produced the momentous outcome that it did, note how easy it is to imagine it having gone the other way and then imagine what we would all have been feeling now. For those whose lives

---

[253] Remember, incidentally, in both the previous US election, and the last UK election, which turned mostly on Brexit, it was actually the preference of a minority of the populations at large that was represented by the electoral majorities.

[254] And at least one author of this book argued that the deaths caused by coronavirus (COVID-19) were a small price to pay for the survival of a truth-based democracy in comparison to the deaths needed to rescue democracy from fascism in the Second World War.

[255] This material has been produced by the V-dem (Varieties of Democracy) Institute an account authored by Christopher Ingraham is published in the *Washington Post* of 12 November, 2020, under the heading 'Democracy Dies in Darkness'. https://t.co/CtT8Ze35l1?amp=1.

turn on the existence of a government-supported professional class tasked with generating truth, it would have meant something close to the end; for everyone who values truth over power, it would have meant despair.

At this point in the immediate post-election days, Trump repeated the charge that he had been rehearsing over several weeks, that postal ballots were subject to fraud and should be discounted. And he set in motion legal challenges to the counting of postal ballots beyond polling day. All this had been anticipated by commentators should the result turn out to be close. But there was also something about it that made the problem we are discussing here, the problem of truth, still more visceral. Trump had no evidence that postal ballots were subject to fraud but, knowing they were more likely to favour Democrats than Republicans, he pronounced on it forcefully, proclaiming that his election victory was being stolen by continued counting of ballots post-election-day. Philosophically, this claim, one of the last he would make in his first term of office, reflected the first claim he made in that term of office, namely that his inaugural crowd had been bigger than Obama's. Both were attacks on the very nature and locus of truth, equally startling and equally penetrating because of their absence of shame or hypocrisy: 'hypocrisy is a tribute vice pays to virtue.'[256] Trump's lies are not disguised, as were the lies of, say, US President Nixon, because they are designed to exhibit the worthlessness of truth in comparison to political power, whereas hypocrisy exhibits the value of truth even as the lie is uttered. Trump has always tried to show that his ability to achieve political success via lying renders truth otiose, or, to look at it another way, it shifts the locus of what counts as truth from truth tellers to the politically potent: to do this one must lie blatantly not furtively.[257]

The authors of this book are academics, belonging to that professional class that values truth like the scientists upon whom much of the argument is modelled. Of course, there are always individual academics for whom power is more important, but the form of life of both science, where it is clearest, and academe in general, still has truth as the defining goal. Now we could see even more clearly that truth is felt as well as heard. Blatant lies are experienced by those who have chosen truth as a calling, as not just wrong, but sensually revolting. It is not just the substance of what is said by truth tellers and liars that is at stake, but its *savour*. To live in *a culture* where truth is the norm is to live

---

[256] François de La Rochefoucauld, 17th century.

[257] To lie *as a principle of government* means trying to destroy the very idea of truth. Hannah Arendt says this is the basis of totalitarianism:

> The ideal subject of totalitarian rule is not the convinced Nazi or the convinced Communist, but people for whom the distinction between fact and fiction (i.e., the reality of experience) and the distinction between true and false (i.e., the standards of thought) no longer exist. (Arendt 1951: 474)

in a warm and sunny land, whereas to live with lies is to freeze in bitter winds. Truth is birdsong, lies are fingernails on a blackboard. It is the warm and sunny birdsong-filled climate that we were hoping to preserve, but a nasty and brutish world that we seemed to be entering.[258] It is because the future is still in the balance that the new US government has to move fast in re-educating the 50%, explaining that the notion that individuals depend on society to be human and that the safety and efficiency of life depends on government, is as true a fact as that diseases are caused by germs not witches.

To make the point again, science is the book's icon because science is the institution, par excellence, which celebrates truth, and which would make no sense as an institution without truth. Science is a check and balance in pluralist democracies, but it is still more important as a champion and exemplar of the love of truth, a truth so precious that it must be guarded by a continual testing of its substance in selfless and even self-destructive ways, ways that allow for no power or opinion or interest to bear on the outcome that owes allegiance to anything but truth itself.[259]

That is why this postscript might as well have been finished on election night. The message was as clear then as it is now: a truth-based democracy is not safe if 50% of the electorate do not feel truth to be a desirable feature of life and do not understand its centrality to democracy. As it happens, the flap of the butterfly's wing has given us a brief moment to rescue proper democracy but if that rescue act is to be substantial, culture will have to change and the true meaning of democracy, and of science, will have to be understood by academics and the citizenry. It will be an uphill battle but every one of our utterances has a part to play.

---

[258] We are not the only people to see it this way. Referring to Trump's exhibition of lying about electoral fraud the comedian and political commentator Stephen Colbert (https://www.youtube.com/watch?v=TeSiJmLoJd0 – 2 minutes 15 seconds in) remarked, 'What I didn't realize is that it would hurt so much. I didn't expect this to break my heart.'

[259] That is science as a form of life; of course there are many instances of individual or even groups of scientists violating the love of the truth for the sake of self-interest, the actions of Andrew Wakefield seemingly being one example, but the institution as a whole cannot change or it would cease to be science. An example of a 'corrupt science' is Lysenkoism but really Lysenkoism was not science at all – 'corrupt science' is an oxymoron.

Appendixes

APPENDIX I

Propaganda and Other Traditions

Any form of widely available remote communication presents the danger of the spread of false information that will be impactful; this has long been realized. We are going to describe briefly the context of this view and go on to suggest that the internet, and social media in particular, presents special dangers.

To go back to the general context, there is a huge advertising industry, and advertisers would not spend the fortunes they do spend if they did not believe that they could influence their audience in meaningful ways – meaningful enough to persuade consumers to spend enough of their own money to provide a handsome return on the cost of the advertisements. The example of tobacco advertising tells us that, at least sometimes, false information will work just as well as true information. The political potential has long been realized. Thus, towards the end of the 19th century fears arose that mass circulation newspapers could influence mass opinion and culture in a negative way. In the first half of the 20th century radio, television and cinema were all cited as mass media technologies with the potential to disrupt democracy.[260] As for conspiracy theories, they have been around for a very long time.[261] Vance Packard's 1957 work *The Hidden Persuaders* looked at the propagandistic influence of television advertising, highlighting the way advertising played upon the emotional

---

[260] Jowett and O'Donnell (2018).
[261] For the history of 'conspiracism' see https://www.the-tls.co.uk/articles/public /modernity-conspiracy-theory-jill-lepore/.

responses of television audiences with the potential to lead them to buy products or support certain political views.[262]

> What [each major form of mass communication that emerged in the 19th and 20th centuries] had in common was their ability to establish direct contact with the public in such a manner as to bypass the traditional socializing institutions, such as the church, the school the family and the political system. (Jowett and O'Donnell 2018, p. 114)

It was realized that mass media content could be presented in such a way as to make audience members feel as if they were being individually targeted, creating a sense of familiarity, or intimacy, between audience member and broadcaster/performer and anticipating what we are arguing in respect of the illusion of intimacy generated by social media. Historically, talk radio generated an identifiable 'radio voice' to maximize the illusion. One 1950s performer suggested that he 'tried to talk to the listener as an individual, to make each listener feel that he knew me and I knew him.'[263]

In the 1930s and 1940s US President Franklin Delano Roosevelt's (FDR) 'fireside chats', transmitted by radio, can be thought of as a benign version of the Big Brother idea – making the President seem like a friend of the family. These broadcasts were directed at both domestic and international audiences and intended to have a political impact from supporting the New Deal and military intervention in the Second World War. Roosevelt imagined the audience for his fireside chats 'in a family group ... sitting on a suburban porch after supper on a summer evening ... gathered around the dinner table at a family meal.' Audience members described him as a 'friend next-door ... and a real fellow who did not talk down to the public.'[264]

Today, radio producers use the same techniques:

> One thing with radio is, you're always told as a producer and as a presenter that it's a one-to-one experience, you're not broadcasting to a crowd of people and saying 'hey all you listeners', it's not like that ... [as a member of the audience] you're being talked to as if you were a single person, so that's what you're trying to get in a programme, that you're talking to one person rather than broadcasting to thousands.[265]

---

[262] The parallels with the fears expressed around social media manipulation firms (see note 235) are striking.

[263] Ryfe (1999, p. 88).

[264] Ryfe (1999, p. 90).

[265] Interview conducted by Mason-Wilkes in the course of his PhD project (Mason-Wilkes 2018).

The effectiveness of this strategy, and the extent to which audience members today are 'taken in' by mass media producers' attempts at cultivating intimacy, are debated. Jowell and O'Donnell claim in their 2018 book that audiences have grown savvy to mass media broadcasters' attempts to cultivate intimacy in this way, and that this awareness allows audiences to resist much mass media propaganda:

> If consumers are aware they are being propagandized, the choice to accept or reject the message is theirs alone. (p. 167)

In older forms of mass media, however effective it is at creating and/or utilizing an illusion of intimacy, some fundamental distance between mass media and any individual member of the mass culture remains, and this is always going to be recognisable to anyone who fails to suspend their disbelief for only a moment: first, mass media communication is one-way only, not two-way with genuinely intimate interaction; second, there will be some visible organization which originates and manages communication via mass media – well-known radio or television stations or long-established newspapers; third, the right to broadcast is carefully guarded and limited, mainly to important or entertaining persons. The illusion of intimacy generated by the internet is not subject to these restrictions.

Of course, as we have mentioned, there are many ways in which face-to-face communication leads to undesirable ends too. For instance, face to face is usually the medium of confidence tricks, of cults and conspiracy theories, and the mass rally allows charismatic individuals to gain political support, sometimes to very bad effect. But there are new forms of remote communication which seem to have special qualities that lend themselves to providing misinformation and disinformation as readily as information. Based on our exploration of the face to face, we want to provide an additional explanation of why this is.

A lot of the reasons for the impact of the internet are already understood. We know that it is now possible to send to send customized and targeted messages cheaply and directly to the smartphones of millions of individuals, without needing the vast resources that would be needed for doing the same kind of thing face to face. It is possible to influence people without needing the skills of the confidence trickster; without the time and trouble needed to build a local cult, with its need for basic living accommodation and extensive indoctrination and which will impact relatively few people anyway; without the organizational skill and apparatus needed for mass rallies and the huge expense for the organizers and the willingness of the participants to travel and to spend their time that way. Even the traditional mass media needs a technical apparatus paid for by taxation or advertisers and it needs highly paid writers, editors and presenters. But anyone can present material on the internet for nothing beyond the cost of their smartphone or wifi subscription. We know that because people use their smartphones the way they do, their personal details can be harvested or hacked by central agencies so that the messages sent to them can be targeted to appeal to them in particular – almost individual by individual. The pervasiveness of the approach has led to it being given its own name, 'surveillance capitalism', its societally and politically distorting dangers being explored in a bestselling academic book.[266] Furthermore, when it comes to elections, it is now possible to identify the uncommitted or floating voters with more certainty and economy than ever before and target them with customized campaigns thus maximizing the value of campaign resources. And we know that targeting individuals is only part of the problem, the other part being that because of the vast number of individuals whose devices can be reached in an organized way, the very background of taken-for-granted reality upon which we rely can be manipulated and shifted so long as the users are ready to interact with their devices and be influenced by what they find there.[267] And we know that what they find there is likely to influence them, even though there is no body language and no commensality, because with the numbers involved and the density of input, what they find now contributes significantly to the bath of words – the sounds and silences – that contribute to the processes of socialization that constitute our way of being in the world.[268]

This is a further influence pushing our reality into new shapes: our reality, the cognitive ground upon which we must push if we are to move forward, is becoming fluid, and if things go on in the same way there will be nothing on which to gain a purchase for argument – the soil of the taken-for-granted, soil which, since the 1970s we know is more like sand than rock, is turning to quicksand; sand provides some purchase if you tread carefully, quicksand provides none.

Social movements

Quite opposite to the main thrust of the argument of this book, at one time social media appeared to hold the promise of being a driver of democratization in authoritarian societies. In *Twitter and tear gas: The power and fragility*

---

[266] Zuboff (2019).

[267] For accounts of how this works see the Netflix movie/documentary *The Big Hack*, or Pomerantsev (2019a) or Zuboff (2019) *Surveillance Capitalism*. Further information can also be found in the recently published insider accounts of Cambridge Analytica by Kaiser (2019) *Targetted* and Wiley (2019) *Mindf*cked*.

[268] UK data from 11,872 adolescents aged 13–15 years show 33.7% reported use of social media less than 1 hour per day (n = 3986); 31.6% reported 1–3 hour average social media use of 1 to 3 hours per day (n = 3720; 13.9% reported 3–5hrs per day (n = 1602); and 20.8%: reported more than 5 hours use per day (n = 2203). (https://psyarxiv.com/z7kpf/).

of networked protest, Zeynep Tufecki investigates the role different network technologies and platforms can and do play in the formation and maintenance of protest movements. Drawing on the concept of technological 'affordances', Tufecki argues that digital social media platforms (Facebook, Twitter, WhatsApp) allowed for the rapid and large-scale mobilization of protestors during, for instance, the Arab Spring and Gezi Park protests in Istanbul. This was in part due to the novelty of these platforms, and thus their relative obscurity to the eyes of repressive regime censors, but also because of the specific 'affordances' of these platforms to diffuse information (protest locations, times etc.) quickly through 'weak ties'. Tufecki further argues that 'digital networking' affords non-hierarchical, 'flat' governance structures, which are particular favoured by movements against authoritarian regimes.

Tufecki, however, argues that though digital platforms allow for the rapid emergence and growth of protest movements with 'flat' structures, this has consequences for the 'staying power' and 'agility' of these movements (thus the *Fragility* in the book's title). Her argument is summarized here:

> 'For example, the ability to use digital tools to rapidly amass large numbers of protestors with a common goal empowers movements. Once this large group is formed, however, it struggles because it has sidestepped some of the traditional tasks of organising. Besides taking care of tasks, *the drudgery of traditional organising helps create collective decision-making capabilities, sometimes through formal and informal leadership structures, and builds collective capacities among movement participants through shared experience and tribulation.* The expressive, often humorous style of networked protests attracts many participants and thrives both online and offline, but *movements falter in the long term unless they create the capacity to navigate the inevitable challenges.'* (Introduction, xxiii, emphasis added)

'Collective decision-making capabilities' and a movement's ability to react tactically and strategically can be thought of as forms of expertise, which require a small, co-located social group in order to develop. Shared experiences and tribulations also increase the solidarity between group members, strengthening group ties and making the group more 'groupish'. In both cases, Tufecki shows face-to-face interaction over an extended period of time is required in order for expertise, solidarity and groupishness to develop, and without the development of these properties, protest movements remain 'fragile'; less coherent, less able to respond to changes in opposition tactics and ultimately easier for regimes to resist.

Jennifer Earl and Katrina Kimport's *Digitally Enabled Social Change* includes similar discussion on the use of web technologies in the development of a range of online social action from 'e-mobilizations' through 'e-tactics' to 'e-movements'. These different kinds of action, which include organising street protests, online petitions, letter-writing campaigns, email campaigns and boy-cotts, 'leverage' the affordances of online technologies (specifically for Earl and Kimport, their relatively low cost and non-reliance on co-presence and co-temporality for organization) to different degrees. Published in 2011, this analysis largely pre-dates the rise of digital social network platforms, and their impacts on the organization of social movements. Earl and Kimport also largely downplay the important role, highlighted by Tufecki, that offline, face-to-face interaction plays in sustaining social movements in the long term, going so far as to suggest that 'collectivity at the level of organizers might not even be neces-sary with some smart uses of the web'. Tufecki's more recent analysis appears to undermine this claim. Unfortunately, it does seem that social media is not the source of long-term changes in authoritarian societies after all.

(i) Coronavirus (COVID-19) Disinformation, (ii) Update on Disinformation in General, and (iii) a Warning about How *Not* to Fix the Problem[269]

Covid, disinformation and misinformation

Ibuprofen

The coronavirus pandemic has also revealed social media's vulnerability to being used to spread false and misleading information. One of the principal challenges for both decision-makers and members of the public has been how to navigate and negotiate a deluge of misinformation and disinformation about the causes and consequences of COVID-19. Some of this material has been malicious, where the rest has been more unwitting, but the key point is that social media and associated platforms have a capacity and capability to distribute such false and misleading messages at a scale and pace that would be unimaginable, where face-to-face communication is the principal mode of information

---

[269] This appendix was first drafted by Martin Innes.

exchange and interaction. Once more the global coronavirus pandemic acts as a living natural experiment in respect of the themes of this book.

An example of the problem was a claim that Ibuprofen should not be used to treat symptoms of COVID-19. In mid-March 2020 claims started to circulate that Ibuprofen and other non-steroidal anti-inflammatory drugs (NSAIDs) should be avoided in the management of COVID-19 symptoms (fever) in favour of Paracetamol (Acetaminophen). It was said that Ibuprofen can: aggravate infection because it accelerates multiplication of the virus; increase mortality risk; and account for the high fatality rate in Italy. Quickly, pharmacies around the world began reporting severe shortages of Paracetamol, with images distributed by social and media sources confirming this and worsening the problem.

Importantly, there may be a 'kernel of truth' to these claims. There is active scientific debate over the use of NSAID drugs like Ibuprofen and Cortisone because their anti-inflammatory action may impact immune system response. There is also scientific inquiry into the role of ACE2 receptors[270] and respiratory disease, as reported for SARS. That said, a scientific consensus would emerge subsequently that there were *no* escalated risks associated with Ibuprofen use.

In terms of understanding how and why this misinformation episode was able to induce a behavioural effect, in the form of the public buying Paracetamol and not Ibuprofen to the point of creating a shortage, it is interesting to trace origins of the panic. The origin of the story appears to be a letter published in the prestigious medical journal *The Lancet* (intriguingly, also the principal source in the case of the MMR revolt). The letter entitled: 'Are patients with hypertension and diabetes mellitus at increased risk for COVID-19 infection?' included the following:[271]

> Human pathogenic coronaviruses (severe acute respiratory syndrome coronavirus [SARS-CoV] and SARS-CoV-2) bind to their target cells through angiotensin-converting enzyme 2 (ACE2) …The expression of ACE2 is substantially increased in patients with type 1 or type 2 diabetes, who are treated with ACE inhibitors and angiotensin II type-I receptor blockers (ARBs). Hypertension is also treated with ACE inhibitors and ARBs, which results in an upregulation of ACE2.5 ACE2 can also be increased by thiazolidinediones and ibuprofen.

Whilst *The Lancet* correspondence contained legitimate scientific questions, its subsequent social media reporting as 'facts' about Ibuprofen risks were misleading because it is:

---

[270] https://www.bmj.com/content/368/bmj.m810/rr-20.

[271] https://www.thelancet.com/journals/lanres/article/PIIS2213-2600 (20)30116-8/fulltext?fbclid=IwAR0ca0qW4HNM7Bbq6fLE1x3L5zP1bsO krP4GXDZ9sO_sm1eV-G8AvamW5fE.

Olivier Véran ✓
@olivierveran

! #COVID – 19 | La prise d'anti-inflammatoires (ibuprofène, cortisone, ...) pourrait être un facteur d'aggravation de l'infection. En cas de fièvre, prenez du paracétamol.
Si vous êtes déjà sous anti-inflammatoires ou en cas de doute, demandez conseil à votre médecin.

10.38 am · 14 Mar 2020 · Twitter for iPhone

43.4K Retweets   **40.4K** Likes

(a) a scientific hypothesis in a letter, not a peer-reviewed research article; (b) discussing long-term NSAID use in specific patient populations; (c) not evidenced in relation to the novel coronavirus.

Two days after the letter's publication, the URL was posted via Facebook to a Spanish medical page *Área Blanca*, in a discussion of co-morbidities and did not mention Ibuprofen. Between 13/03/20 and 15/03/20, there was a steady growth in article shares on Facebook before appearing at high volumes on Twitter on 16 March (N=149 shares), the same day that Facebook shares peaked at 42. In the process the content of the letter was being misinterpreted and blended with information deriving from other sources, increasing the extent of the misinformation. Most notable in this regard were a series of messages distributed widely on WhatsApp, using multiple 'spoofed' medical doctor personas, as well as a number of blogs in a variety of languages.

The 'reach' of the misinformation was substantially boosted when the French Health Minister Olivier Veran, who is also a medical doctor, featured it in a tweet on 14 March.

Translation: #COVID – 19 | Taking anti-inflammatory drugs (ibuprofen, cortisone, ...) could be a factor in worsening the infection. If you have a fever, take paracetamol. If you are already on anti-inflammatory drugs or in doubt, ask your doctor for advice.

As a French Minister, qualified doctor and neurologist, Véran was a highly credible messenger, with his message receiving more than 43k RTs and 40k Likes. The next day, Bulgarian news agency Novinite wrote Véran's account had been hacked, although this claim was subsequently revised.[272] In the UK,

---

[272] https://www.novinite.com/articles/203622/Fake+News:+Ibuprofen+and+Cortisone+may+Worsen+your+Condition+if+you+are+Infected+with+COVID-19.

Véran's tweet was amplified by a Guardian article (500 shares to Facebook pages; Twitter = 55, Reddit = 45; Instagram = 2).[273]

Covid-19 origin conspiracies

A large number of additional misleading narratives have gravitated around the coronavirus, influencing public understandings and interpretations. For example, by 15 March 2020, which was still quite early in the pandemic, the Cardiff team had counted at least 59 distinct conspiracy theories and disinformation narratives concerning the causes and consequences of Covid-19. Whilst some of these were new, many 'reheated' and updated long-standing conspiracies. For example, one that gained significant traction in countries across Europe was the idea that the transmission of the virus was connected to the new 5G mobile phone network. In the UK, the anger and concern that this generated was connected to over 70 physical attacks, including arson, on phone masts in certain areas of the country.

Other conspiracies about the origins of coronavirus included the idea that it was connected to a scientific research programme in Wuhan, China, that was especially popular amongst 'hard-right' American 'patriot' online communities, circulating in high volumes across the alternative media ecosystem associated with such groups. There was also a counter-conspiracy that the virus was a bio-weapon engineered by the US military in Fort Detrick, that was pushed by Russian and Chinese state media sources and their affiliates.

Face masks

A further example of how science can become enrolled in the propagation of misinformation about coronavirus (COVID-19) can be found in the resistance that was generated to policies requiring citizens to wear face masks. Specifically, across a number of established online communities that were highly sceptical of the coronavirus threat and maintained general anti-vaccine type viewpoints, a claim started circulating that regular mask wearing heightened the risks of CO_2 toxicity (aka hypercapnia).

One advocate of the hypercapnia thesis had published his ideas, on the platform ResearchGate, formatted to look like a genuine scientific report. This is part of a wider emergent pattern where highly controversial research is being 'published', prior to any proper peer-review in legitimate scientific fora, to manipulate its apparent credibility and gain publicity. The UK group 'Lockdown

---

[273] https://www.theguardian.com/world/2020/mar/14/anti-inflammatory-drugs -may-aggravate-coronavirus-infection (accessed 7 June 2020).

Sceptics'[274] and the Daily Mail both repeated the respective dangers of CO_2 toxicity and the health dangers of exercising with a mask.[275]

Update on disinformation in general

Three years on from the original 'discovery' of the Kremlin's information interference and influence operation, the situation looks worse. Over the past two years, Twitter has released datasets listing accounts that they consider to have engaged in 'coordinated inauthentic behaviour'. Table A2.1 summarizes some of the material that has recently been released.

As can be seen, the number of states involved and the number of accounts has increased hugely. It is likely that these are only the more poorly disguised activities. Hence our growing concern that the distribution of disinformation is in danger of becoming a 'normalized' feature of contemporary political and social life.

Table A2.1: Twitter Accounts Recently Identified as Engaging in 'Coordinated Inauthentic Behaviour' by Alleged Country of Origin.

COUNTRY OF ORIGIN	NO. OF ACCOUNTS	TWITTER RELEASE DATE
China	5241	July 2019
Ecuador	1019	April 2019
Iran	770	October 2018
	2320	January 2019
	4779	June 2019
Russia	3613	October 2018
	416	January 2019
	4	June 2019
Saudi Arabia	6	April 2019
Spain	259	April 2019
United Arab Emirates	4248	March 2019
(+ Egypt)	271	April 2019
Venezuela	1960	January 2019

---

[274] https://lockdownsceptics.org/.
[275] https://www.dailymail.co.uk/news/article-8311179/Joggers-lung-collapses-ran-2-5-miles-wearing-face-mask.html (accessed 12 May 2021).

A warning about how *not* to fix the problem

Currently, 'technical fixes' exhaust the proposals for mitigating the problems whereas we have argued in this book that a change in the taken-for-granted and a halt to the process of normalization of the liquification of knowledge are what is needed – a far more difficult prospect. Among the proposed technical fixes, fact-checking methodologies are a favourite but they have little purchase when set against the influence and impact of a well-crafted disinformation narrative. The distorting and deceptive message is likely to have acquired traction before any fact-checking mechanism can disrupt its transmission and reception.

And, of course, most of the decision-making power when it comes to responding to disinforming communications resides with the commercial companies who run the social media platforms and who have an interest in maintaining their self-policing status. Self-policing ensures that commercial concerns will play a large part in the decisions. At the same time, governmental regulatory apparatuses are always troubled by the problem of balancing interventions with commitments to freedom of expression, and the transnational dimension of the contemporary media ecosystem. It is unsurprising, then, that the problem is getting worse, as documented in Table A2.1. Increasing polarization in 'the Western democracies' driven by processes such as the Brexit campaign, and the US President's daily insults against reality, mean that the governments of countries that should be most resistant to the spread of disinformation so that they can preserve pluralist democracy have discovered an interest in fostering it and normalizing it irrespective of the long-term consequences for their way of life. This is one of the things that has to change if pluralist democracies are to survive. We have argued that the institution of science – seen as craftwork with integrity – may have a role to play.

Ironically, some of the well-intentioned *ethical* decisions taken by social scientists about how to study social media and its implications are accentuating the problem. Social science research is in danger of being unwittingly complicit in the production and reproduction of disinformation:

In an effort to configure ethical principles for working with social media data that are compatible with conventions for offline research, various learned societies and individual scholars have put together ethical guidelines for conducting digital social science on behalf of their members and colleagues. One feature is the rule that if social media users delete posts from their timelines, scholars reporting on their activity should not make reference to the removed data. One can understand that the motive for this idea is positive – rather as when it is agreed that it should be possible for interviewees in social science projects to be able to reconsider any quotations in case they felt they spoke hastily or infelicitously – but this simply does not work in the case of social media research since deletions are an integral part of the data that is being researched.[276] If this

---

[276] For an argument for allowing respondents in recorded interviews to edit their quotations, see Collins et al. (2019).

principle is followed, then researchers could find themselves misrepresenting the record of how certain events unfolded.

For example, in the immediate wake of the Westminster Bridge terror attack in London in 2017, a social media account positioned as specializing in 'breaking news' tweeted a picture from the scene of the alleged assailant, alongside an image of well-known Islamist extremist Abu Izzadeen, claiming that they were the same person (there was some physical resemblance between the images). Very shortly afterwards, a French and an Italian news organization used their social media accounts to repeat the allegation. About 45 minutes later, an unknown individual edited Izzadeen's Wikipedia page, inserting the claim that he was responsible for the terror attack. Broadcasting live from the scene, Channel 4 News then opened their programme, with the presenter repeating the claim originating on social media that a suspect for the attack had been unofficially identified as Abu Izzadeen.

Shortly after this, doubts began to surface about the veracity of the identification. Indeed, it was debunked when it was revealed that Izzadeen was currently serving a prison sentence and so could not have been responsible. Channel 4 News closed that evening with a retraction of the claim.

Two consequences of this episode are pertinent to the concerns of this book. First, despite it being rapidly and thoroughly falsified, the idea that a well-known Islamist extremist was responsible for the Westminster attack continued to circulate and was shared for some considerable time afterwards on a number of extreme far-right online forums. The denial was described as possible evidence of a 'deep state' led 'false flag' conspiracy. At some point that evening, both the Italian and French media outlets sought to delete any digital traces that they had spread an incorrect story from their social media profiles, suggesting instead that it was Channel 4 News and the *Independent* newspaper that were responsible for the rumour. Thus, if we were to follow the ethical guidelines described above, this would involve being complicit in the construction and communication of a false story about how this piece of disinformation was promulgated.

The problem is exemplified more widely where individuals who commit hate crimes on social media, or express racist or homophobic views, then seek to cover their tracks by expunging them from their digital record. Moreover, this approach of dropping contentious and emotionally loaded messages into the social media stream, and then deleting the originating source, is a specific technique honed by significant disinformation actors such as the Russian IRA.

Craftwork with integrity in social science cannot abandon the 'integrity' part, yet complicity in concealing the fact that certain postings were subsequently deleted would be doing precisely this. Here there are cross-cutting moral imperatives – a well-meaning attempt to safeguard the rights of internet users amounts to a sacrifice of integrity in social media research to the extent of complicity in the aims of certain dangerous groups. Given the current state of the world and the assault on pluralist democracies by disinformation campaigns, it is the latter harm that seems incomparably greater.

The Delineated Cases of Citizen Participation in Science and Technology

The 'delineated cases' of the interaction of citizens and technical specialists are those where citizens or other groups have a legitimate rather than 'magical' contribution to make to specialist debate; there are three kinds, often not distinguished by those in favour of the democratization of science. The three types are: (i) cases where citizens have legitimate rights in virtue of some relevant, specialist experience; (ii) cases where their participation is justified by more ubiquitous experiences; and (iii) cases where the dispute is really about how to frame the question in the first place. In the first type of case, the rationale for including citizens is epistemic; in the other two, the contribution of citizens lies in their ability to represent different sets of values. The 'democratization' approach obscures these differences just as it pays no attention to the difference between the delineated cases and citizens' role in elite technical expertise in general.[277]

In the first type the so-called 'citizens' (or 'lay experts') are really experts – experience-based experts – who have developed their own bodies of specialist

---

[277] See also Evans (2011); Evans and Kotchetkova (2009); Evans and Plows (2007). Susskind and Field (1996) discuss such 'delineated cases'.

knowledge. Although not scientific knowledge, this expertise is as rich in tacit knowledge as scientific knowledge and can be a vital component of decision-making. That it is not valued tells us more about the way societies value different communities than the kinds of expertise they have to offer.[278] These ideas are clearly illustrated in the well-known instances of the Cumbrian sheep farmers and the farm-workers responsible for spraying the organo-phosphate 245-T. In these cases, widening participation does not need to be justified on democratic grounds as the excluded groups have relevant expertise and should be contributing as experts in their own right. This requires some changes to the ways governments think about the sources of expertise they listen to when seeking advice, but such changes are more accurately seen as improving expert advice rather than democratizing science.

The second type arises when a value judgement among stakeholders is required to choose between different options for action. Examples of such decisions include planning disputes where a new development might create more jobs but only at the cost of more pollution. Here experts might be able to describe the trade-off but they should not be asked to choose between different outcomes as that is a political judgement not a scientific or technical one. Whilst it is usually the case that political judgement are best made by those most directly affected, and local communities have the most direct experience of the conduct of the institutions in their area, this kind of local discrimination will need to be balanced against competing regional and national priorities.[279] Such a balancing act is a political choice, not a technical choice, but history shows that it is too often the case that purely technical experts do make the decision, compounding the problems. It is important to note, however, that even if all the relevant experts had been consulted, the choice that remains would, in the last resort, be a political/policy choice, and so the challenge here is to improve democratic systems so that citizen views are adequately represented and heard.[280]

---

[278] For more on the role of tacit knowledge in the development of expertise see Collins and Evans (2007).

[279] 'Local discrimination' is a term taken from Collins and Evans (2007) – see especially the 'Periodic Table of Expertises' – and refers to the ability of local communities to use more informed judgements of trust to choose between competing expert claims. Whilst not technical judgements, and having no influence on what technical experts believe, for the citizens involved choosing *who* to believe also resolves the problem of *what* to believe to be true, giving rise to the idea that social judgements can be 'transmuted' into technical ones.

[280] Susskind and Field (1996) discuss such cases and describe a 'mutual gains' approach that can be used to maximise the chances of successful resolution.

The third type occurs when the nature or scale of the question being asked is itself the source of the controversy. Whilst often seen as a 'technical' issue, the framing of a problem is really a value-laden choice about the 'costs' and 'benefits' that need to be weighed in any political decision and the kinds of experience and evidence are needed establish their significance. This leads to the argument that the process of soliciting and using expert advice should be democratized by so-called 'upstream engagement' which aims to include a wide range of perspectives at the beginning in order to avoid prematurely narrowing the debate; here it is being argued, usually implicitly, that citizens' ubiquitous expertises should be engaged since the aim is to explore political, technical and moral possibilities in order to identify concerns and priorities in a more collective way. Note that this kind of delineated case still does not extend to the technical debate which comes once the overall parameters have been set.

An Alternative View: Successful Business Interaction *Without* Face-to-Face Communication

Within the organization studies literature there are published studies which suggest that remote interaction can work just as well in business transactions without the need for local trust to be developed. A comprehensive meta-analysis published in 2014 looked at the effectiveness of virtual teams – teams that communicate only remotely.[281] The analysis finds that experimental research with students simulating virtual teams shows them to be less effective than local teams, whereas field studies reveal a number of examples of extraordinarily effective virtual collaborations which resulted in clear efficiency and financial gains for the involved firms. One of the most challenging in terms or our analysis of the importance of face-to-face communication is the multi-organization development of a new rocket engine – known by the acronym 'SLICE' – by Boeing and collaborating firms which, it is said, was developed in an unprecedentedly short time, with huge reliability and cost savings, all through remote communications.[282] The paper describing the collaboration explains the conditions required for such an effective collaboration, including multiple

---

[281] Purvanova (2014).
[282] Malhotra et al. (2001).

teleconferences. In other words, as much as possible in the way of the simultaneity of face to face was recreated even though communication was remote. We should also note that studies of this kind, if they are to bear on our problem, must exercise the most careful monitoring and 'hygiene' in respect of what is going on in the face to face. If members succeed in managing a project remotely but only after trust has been acquired and tacit understanding has been transmitted by earlier small group interactions, then we need to know. We need to know because we already know that remote interaction is effective once these things have been managed. In terms of Table 10.1, we need to know if what is being described is really the ultimate column – the equivalent of fake crab – or is it some kind of partial version of regular communication with a heavy stress on remote interaction. There is no reason to think, given their purpose, that the studies included in the meta-analysis would have included a high degree of hygiene and monitoring, since their project was different to ours.

Another difficulty with the SLICE report is that there is no detail of how new understandings were developed or misunderstandings resolved and, as a result, we do not really know how hard the problem was from the outset. The authors are very clearly aware of what we can call the 'incommensurability problem' – the difficulty arising out of different 'languages' being spoken by team members being drawn from different communities who frame problems and perceive the world in different ways (see Chapter 2, Duck-rabbits and the bath of words for an illustration of this problem), but the evidence they provide for the problem being present and resolved in the SLICE case study is simply the different *organizational* origin of the contributing members, not the different cognitive communities. They explain that the members came from different firms with different knowledge-sharing traditions and that they came from different disciplinary backgrounds. But, on the other hand, what was being done was to develop an improved rocket engine with rocket engines being decades old at the time, and the disciplines, though a mix, were all within the umbrella of rocket engineering. As the 20th century moved into its last quarter, 'rocket science', though still often used as a paradigm of brilliant thinking as in 'it is not rocket science', began to be found in the domain of the conceptually familiar even if technically demanding. If we consider the mix of engineers contributing to the SLICE engine we are, at best, unsure of what kind of contribution they were making in the sense of the notion of contribution discussed in relation to the definition of contributory expertise is discussed in the text (see Chapter 1, Studies of expertise and experience (SEE)). We can again describe the problem in terms of an earlier summary analysis, this time Figure 10.1. The question is where this domain is situated along the X-axis; if it is far to the left there is no problem but if it is far to the right, then the claims clash with the claims of this book. In other words, the crucial questions are (i) the extent to which all members of a collaboration have to understand the entire project or can simply supply discrete expertises to be melded in via the overarching understanding (e.g. via interactional expertise) of a sub-set of leading contributors; and (ii) the extent to which potential understanding of the entire project is likely to have come with the scientific or engineering education that was acquired by every contributor prior to their joining the group – the 'rooting' of the remote in F2F. The more conditions (i) and or (ii) are fulfilled, the less is there likely to be a problem with remote collaboration that needs to be resolved with F2F. The more conditions (i) and (ii) are met, the more does the collective problem come to look like a problem of the *organization* of the supply of goods – in this case intellectual goods – that is, the employment of technical services, in this case, creative technical services; the more, in other words, are we looking at a case of 'trade', which proceeds smoothly, rather than of a problematic 'trading zone', where alternative meanings and incommensurabilities are in play.[283] In the SLICE paper, no attention is paid to these things and all the difficulties are laid at the door of the different organizational cultures – a real problem to be sure but one with a straightforward set of solutions that the authors describe.

In sum, insofar as this highly cited paper on the SLICE case study is representative of the organizational studies literature, it shows that that this kind of analysis of virtual groups cannot be generalized too far because they do not deal with hygiene and monitoring of earlier interactions between the parties, and do not examine in detail the extent to which conditions (i) and (ii) are met. Without that kind of detail, we do not know what kind of problem we are dealing with; this, unfortunately, renders the meta-analyses less informative than they might be, comprehensive and interesting though they are.

Another way of looking at things is represented in Figure A4.1, which is intended to represent, on three axes, what we will call the 'trading space' between groups engaged in a 'trading zone'. A trading zone is the zone of interaction between specialist domains with more or less incommensurable ways of describing and manipulating the world.

The Z-axis, going into the page, represents the extent to which one group simply delivers goods and services to another without engaging with their problems. At the origin of the Z-axis they might deliver engineering solutions to well-specified problems in the same spirit as they might deliver fuel for the central heating. At the far end of the Z-axis, the groups might come to understand each other's worlds and problems, participating in each other's specialist cultures and becoming culturally as like each other as possible in the course of the interaction. The, vertical, Y axis, represents the, related, extent to which the groups acquire each other's tacit knowledge while the horizontal, X-axis, is the extent to which they engage with each other remotely or face to face.

The axes of the trading space are not independent: you cannot go far along the Y axis without going along the Z-axis – you cannot acquire lots of tacit knowledge without engaging. And, of course, the major thesis of the book is that

---

[283] Collins, Evans and Gorman (2007, 2019).

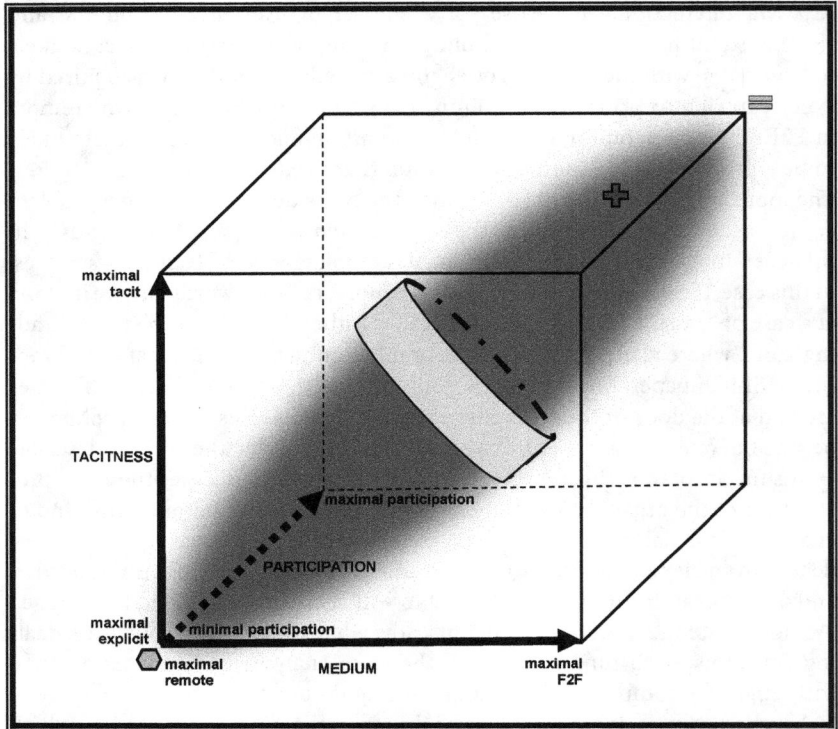

Figure A4.1: The thesis.

you cannot go far along either Z or Y without going along X: you cannot go far in acquiring a group's tacit knowledge or in participating in that group's activities, without engaging in a quite a lot of face-to-face encounters. So the trading space is really an alternative way of *describing* our argument. This is that the positions available in the trading space are limited to those bounded by the fuzzy, diagonal, cigar-like shape. Trust and efficiency too could well be represented in the diagrams too, increasing as you go up the cigar from the position indicated by the hexagon at the bottom-left to that indicated by the equals-sign at the top-right.

Our question about the organizational studies cases is the extent to which they are found toward the origin of the Z-axis. If that is where they are found, it is not surprising that that they are near the origin, in respect of the extent of remote communication and the ratio of explicit to tacit knowledge. Only if the cases were far from the origin, yet still used virtual groups successfully, would a challenge be set for the theses about face-to-face communication developed here: such cases would represent positions outside the boundaries of the 'cigar'.

The reason the problem of cross-cultural trading zones is sometimes not noticed in modern societies, with technical services seemingly melded without any specially developed mutual understanding, is that the cooperating specialist subcultures often lie within an overarching shared technical culture; in such a case the Face-to-Face Principle is satisfied and the overarching culture, born in a foundational process of socialization, provides enough of a common language and understanding to make it possible for groups to cooperate with a minimum of mutual socialization. Here is an expanded version of the quotation from Michael Polanyi which relates to the principle:

> Now we see tacit knowledge opposed to explicit knowledge; but these two are not sharply divided. While tacit knowledge can be possessed by itself, explicit knowledge must rely on being tacitly understood and applied. Hence all knowledge is either tacit or rooted in tacit knowledge. A wholly explicit knowledge is unthinkable. (quoted in *Knowing and Being*, p. 144)[284]

Quite simply, whenever one is, say, reading a set of instructions, the comprehensibility of them depends on things one cannot say – one cannot, for example, describe what skills one is employing to read the words or to understand the words, and one cannot describe how one understands the instructions contained in the words even if one understands the meaning of each word – each rule leads to the need for another rule to explain it.[285]

---

[284] Grene (1969). For the concept of tacit knowledge as it applies to this problem see Collins (2001, 2010).

[285] For an illustration of this point see the game 'Awkward Student' as found in Collins (1985/92, p. 13–15). In this game the awkward student must (and can) find ways not to understand the simple instruction 'continue the series 2, 4, 6, 8'.

Second Language Learning

When it comes to second language learning, Kuhl, who has been discussed earlier, claims there is a sharp drop-off in ability to acquire fluency in a language after the age of seven as shown in Figure A5.1, which is a rough re-drawing of her published graph (Kuhl 2010, p. 716). Of course, what could be being demonstrated here is changes in the plasticity of the brain, but since second languages are often learned later in life, we might expect to discover that fluency was reduced in second languages and, in so far as it was maintained, conditions very similar to that under which fluency was initially acquired would be necessary – that is face-to-face interaction with native speakers.

The matter of accent is an interesting one even before we examine the actual research. What we are looking for when we explore the advantages of F2F is, in some senses, less demanding than acquisition of fluency in accent or even grammatical fluency. Studies of early socialization take an accent that cannot be distinguished from that of (some set of) other natives by (some set of) other natives as a criterion of full enculturation. But *secondary* socialization into some esoteric scientific domain to the extent of proficient interactional expertise can be accomplished with broken grammar and a foreign accent; passing a test of interactional expertise does not require passing as a native-language speaker! It follows that we should not expect to find that secondary socialization that satisfies our criteria of 'cultural fluency' can be accomplished only if it is begun before a certain age – it is less demanding than that. But it is more demanding that acquiring fluent grammar, which computers can do, since it requires an

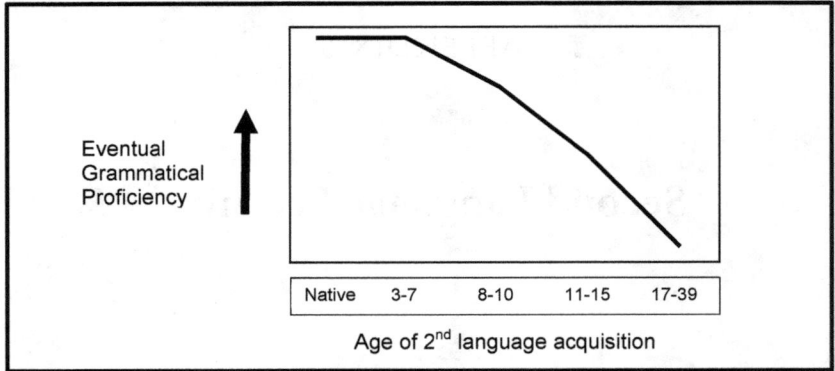

Figure A5.1: Kuhl's model of age of fluent language acquisition approximately represented.

understanding of practices, which computers cannot yet accomplish (as the Winograd schema example shows).

How, then, does work on secondary socialization bear on our concerns? A 2018 study of 2/3 million English speakers looks at the problem of second language learning in general, extending the study to later ages.[286] This was organised as a game with volunteers playing over the internet. They answered questions which revealed their understanding of subtle features of English grammar and usage and filled out a questionnaire that indicated their exposure to the language. At the time of writing, the game can be found at www.gameswith words.org /WhichEnglish. At the end of each game the programme would try to identify the kind of English spoken by the player (e.g. English vs. American English or other dialects). The research project was aimed at discovering if there was an average age beyond which full fluency could no longer be acquired, and what that age was. The conclusion was that there was such an age but that it was late teens – much later than had previously been thought.

For our purposes the most interesting result is expressed in Figure A5.2. The lines show language accuracy on the vertical axis against age of first learning on the horizontal axis, the upper line being for immersion learners and the lower line being for non-immersion learners. Here, immersion learners, of whom there were 45,067, were defined as either simultaneous bilinguals who grew up learning English simultaneously with another language (age of first exposure = 0), or later learners who learned English primarily in an English-speaking setting (defined as spending at least 90% of their life since age of first exposure in an English-speaking country). Non-immersion learners, of whom there were 266,701, had spent, at most, 10% of post-exposure life in an

---

[286] Hartshorne et al. (2018).

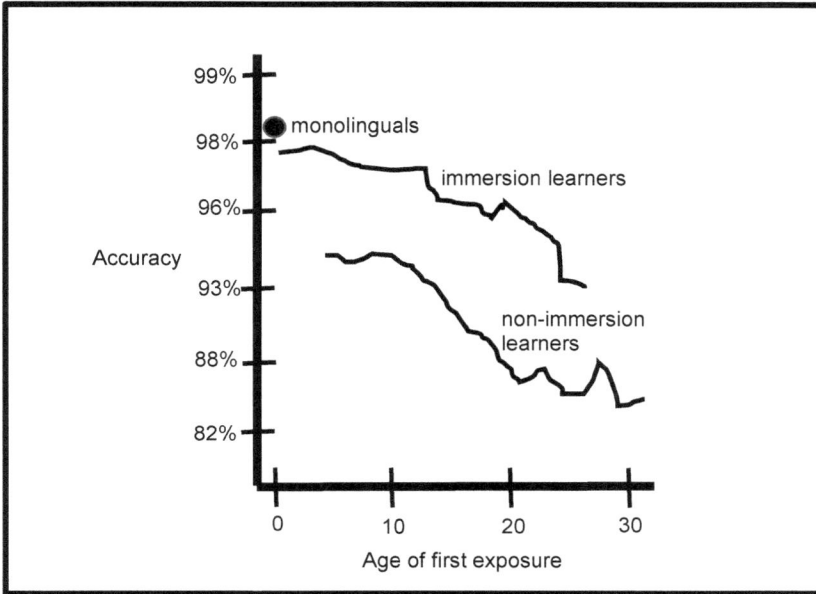

Figure A5.2: Apparent success of immersion learners compared to non-immersion learners in a large study of English learned as a second language (roughly adapted from Hartshorne et al. 2018, Figure 6, p. 270).

English-speaking country and no more than one year in total. These definitions reveal, once more, how relatively undemanding are our criteria of secondary socialization, since these levels of exposure are far beyond what are expected of the most conscientious of sociologists or anthropologists – to count someone who has spent nearly a year of their lives in a target society as a 'non-immersion learner' shows just how different the standards are.

Nevertheless, the figure appears to show, unambiguously, that immersion is better than non-immersion – at every starting age, the immersion learners achieved a notably higher level of accuracy than the non-immersion learners.

Unfortunately, things are not so simple. The problem, as the authors point out, is that immersion and non-immersion are confounded with sheer density of exposure to the language. The relative success of immersion learners might be due not to the subtle qualities of face-to-face interaction with native speakers but to the sheer intensity of exposure to the language under immersion learning as compared to other kinds of learning. Our common sense suggests that something more is involved but this study does not prove it. Furthermore, all the studies reported, other than those that deal with accent, measure success in second language learning by grammatical accuracy, not the more subtle features of fluency which involve understand practices and social contexts.

References Cited

Abbott, Andrew 2001 *Chaos of disciplines*. University of Chicago Press. DOI: https://doi.org/10.7208/chicago/9780226001050.001.0001.

Alkhateeb, Ahmed 2017 Science has outgrown the human mind and its limited capacities. *Aeon* (24/04/2017). Available at: https://aeon.co/ideas/science-has-outgrown-the-human-mind-and-its-limited-capacities [accessed 18/08/2021].

Allport, G W 1954 *The nature of prejudice*. Addison-Wesley.

Arendt, Hannah 1951 *Origins of totalitarianism*. Schocken Books.

Arminen, Ilkka, Segersven, Otto E A and Simonen, Mika 2018 Active and latent social groups and their interactional expertise. *Acta Sociologica*, 62(4): 391–405. DOI: http://dx.doi.org/10.1177/0001699318786361.

Asch, Solomon E 1951 Effects of group pressure on the modification and distortion of judgments. In: Guetzkow, Harold. *Groups, Leadership and Men.* Carnegie Press. pp. 177–190.

Asimov, Isaac 1991 *The naked sun*. Bantam Books (first editions 1956/1957).

Atkinson, Carol 2007 Trust and the psychological contract. *Employee Relations*, 29(3): 227–246. DOI: https://doi.org/10.1108/01425450710741720.

Atkinson, J Maxwell 1984 *Our masters' voices: The language and body language of politics*. Methuen.

Barthes, Roland 1968 La mort de l'auteur [The death of the author]. *Manteia*, 5: 12–17.

Bates, Thomas R 1975 Gramsci and the theory of hegemony. *Journal of the History of Ideas*, 36(2): 351–366. DOI: https://doi.org/10.2307/2708933.

Bauman, Zygmunt 2000 *Liquid modernity.* Polity Press.

Baym, Nancy K 2015 *Personal connections in the digital age.* Polity Press.

BBC News 2018 How the Dutch foiled Russian 'cyber-attack' on OPCW. 4 October. https://www.bbc.co.uk/news/world-europe-45747472 [accessed 08/12/2021].

BBC News 2020 Trump Covid post deleted by Facebook and hidden by Twitter. 6 October. https://www.bbc.com/news/technology-54440662 [accessed 08/12/2021].

Beck, Ulrich 1992 *Risk society: Towards a new modernity.* London; Newbury Park, Calif: Sage Publications.

Bittman, Ladislav 1985 *The KGB and Soviet disinformation: An insider's view.* Pergamon-Brassey's.

Bligh, Donald A 2000 *What's the use of lectures?* Intellect Books.

Block, Fred L and Keller, Matthew R (eds.) 2011 *State of innovation: The U.S. government's role in technology development.* Paradigm Publishers.

Bloor, David 1983 *Wittgenstein: A social theory of knowledge.* Macmillan. DOI: https://doi.org/10.1007/978-1-349-17273-3. PMCid: PMC1916315.

Blume, Stuart 2017 *Immunization: How vaccines became controversial.* Reaktion Books.

Boghardt, Thomas 2009 Soviet bloc intelligence and its AIDS disinformation campaign. *Studies in Intelligence,* 53(4): 1–24.

Bolukbasi, Tolga, Chang, Kai-Wei, Zou, James, Saligrama, Venkatesh and Kalai, Adam 2016 Man is to computer programmer as woman is to homemaker? Debiasing word embeddings. Available at: https://arxiv.org/abs/1607.06520.

Bonikowski, Bart 2017 Ethno-nationalist populism and the mobilization of collective resentment. *British Journal of Sociology,* 68: S181–S213. DOI: https://doi.org/10.1111/1468-4446.12325. PMid: 29114869.

Botsman, Rachel 2017 *Who can you trust? How technology brought us together and why it could drive us apart.* Penguin/Random House.

Bowlby, John M 1953 *Child care and the growth of love.* Penguin.

Boyce, Tammy 2006 Journalism and expertise. *Journalism Studies,* 7(6): 889–906. DOI: http://dx/doi/org/10.1080/14616700600980652.

Boyce, Tammy 2007 *Health, risk and news: The MMR vaccine and the media.* Peter Lang.

Breitbart, Andrew 2011 *Righteous indignation: Excuse me while I save the world.* Grand Central Publishing.

Butler, Judith 1990 *Gender trouble: Feminism and the subversion of identity.* Routledge.

Centola, Damon 2018 *How behaviour spreads: The science of complex contagions.* Princeton University Press. DOI: https://doi.org/10.23943/9781400890095.

Chater, Nick and Christiansen, Morten H 2018 Language acquisition as skill learning. *Current Opinion in Behavioural Science,* 21: 205–208. DOI: https://doi.org/10.1016/j.cobeha.2018.04.001.

Chu, S and Kim, Y 2011 Determinants of consumer engagement in electronic word of mouth (eWOM) in social networking sites. *International Journal of Advertising*, 30(1): 47–75. DOI: https://doi.org/10.2501/IJA-30-1-047 -075.

Cialdini, Robert 2009 *Influence: Science and practice*. William Morrow.

Cohan, William D 2021 My years on Wall Street showed me why you can't make a deal on Zoom. *New York Times*, 16 August. Available at: https://www.nytimes.com/2021/08/16/opinion/covid-wall-street-delta -office.html [accessed 18/08/21].

Colbert, Stephen 2020 Stephen rips up the monologue and starts over after Trump's heartbreaking Thursday Night Lie Fest. *YouTube*, 6 Nov. Available at: https://www.youtube.com/watch?v=TeSiJmLoJd0 [accessed 08/12/ 2021].

Collins, Caity 2020 Productivity in a pandemic. *Science*, 7(369): 603. DOI: https://doi.org/10.1126/science.abe1163. PMid: 32764040.

Collins, Harry 1974 The TEA set: Tacit knowledge and scientific networks. *Science Studies*, 4(2): 165–186. Available at: https://www.jstor.org/stable /284473. DOI: https://doi.org/10.1177/030631277400400203.

Collins, Harry 1984 Concepts and methods of participatory fieldwork. In: Colin Bell and Helen Roberts. *Social Researching*. Routledge. pp. 54–69.

Collins, Harry 1985/92 *Changing order: Replication and induction in scientific practice*. Sage [2nd edition, University of Chicago Press].

Collins, Harry 1992 Hubert L. Dreyfus, forms of life, and a simple test for machine intelligence. *Social Studies of Science*, 22(4): 726–739. DOI: https:// doi.org/10.1177/030631292022004008.

Collins, Harry 1996 Interaction without society? What avatars can't do. In: Stefik, Mark. *Internet Dreams*. MIT Press. pp. 317–326.

Collins, Harry 2001 Tacit knowledge, trust, and the Q of sapphire. *Social Studies of Science*, 31(1): 71–85. DOI: https://doi.org/10.1177/030631201031001004.

Collins, Harry 2004a Interactional expertise as a third kind of knowledge. *Phenomenology and the Cognitive Sciences*, 3(2): 125–143. DOI: https://doi .org/10.1023/B:PHEN.0000040824.89221.1a.

Collins, Harry 2004b *Gravity's shadow: The search for gravitational waves*. University of Chicago Press. DOI: https://doi.org/10.7208/chicago/978022611 3791.001.0001.

Collins, Harry 2010 *Tacit and explicit knowledge*. University of Chicago Press. DOI: https://doi.org/10.7208/chicago/9780226113821.001.0001.

Collins, Harry 2011 Language and practice. *Social Studies of Science*, 41(2): 271–300. DOI: http://dx.doi.org/10.1177/0306312711399665. PMid: 21998924.

Collins, Harry 2014 *Are we all scientific experts now?* Polity Press.

Collins, Harry 2016 An Imitation Game concerning gravitational wave physics (being Chapter 14 of *Gravity's Kiss*). Available at: http://arxiv.org /abs/1607.07373.

Collins, Harry 2017 *Gravity's kiss: The detection of gravitational waves.* MIT Press.

Collins, Harry 2018 *Artifictional intelligence: Against humanity's surrender to computers.* Cambridge, UK: Polity Press.

Collins, Harry 2019 *Forms of life: The method and meaning of sociology.* MIT Press.

Collins, Harry, Barnes, Bill and Sapienza, Riccardo 2020 The dangers of going online only. *Physics World*, 2 July, p. 19. DOI: https://doi.org/10.1088/2058 -7058/33/7/24.

Collins, Harry, Bartlett, Andrew and Reyes-Galindo, Luis 2017 Demarcating fringe science for policy. *Perspectives on Science*, 25(4): 411–438. DOI: https://doi.org/10.1162/POSC_a_00248.

Collins, Harry and Evans, Robert 2002 The third wave of science studies: Studies of expertise and experience. *Social Studies of Science*, 32(2): 235–296. DOI: https://doi.org/10.1177/0306312702032002003.

Collins Harry and Evans, Robert 2007 *Rethinking expertise.* University of Chicago Press. DOI: https://doi.org/10.7208/chicago/9780226113623.001.0001.

Collins, Harry and Evans, Robert 2014 Quantifying the tacit: The Imitation Game and social fluency. *Sociology*, 48(1): 3–19. DOI: http://dx.doi .org/10.1177/0038038512455735.

Collins, Harry and Evans, Robert 2015 Expertise revisited I – Interactional expertise. *Studies in History and Philosophy of Science*, 54(December): 113–123. DOI: https://doi.org/10.1016/j.shpsa.2015.07.004. PMid: 26568093.

Collins, Harry and Evans, Robert 2016 The bearing of studies of expertise and experience (SEE) on ethnography. *Qualitative Inquiry*, 23(6): 445–451. DOI: http://dx.doi.org/10.1177/1077800416673663.

Collins, Harry and Evans, Robert 2017a Probes, surveys and the ontology of the social. *Journal of Mixed Methods Research*, 11(3): 328–341. DOI: http:// dx.doi.org/10.1177/1558689815619825.

Collins, Harry and Evans, Robert 2017b *Why democracies need science.* Polity Press.

Collins, Harry, Evans, Robert, Durant, Darrin and Weinel, Martin 2019 *Experts and the will of the people: Society, populism and science.* Palgrave. DOI: https://doi.org/10.1007/978-3-030-26983-8.

Collins, Harry, Evans, Robert and Gorman, Michael 2007 Trading zones and interactional expertise. In: Collins, Harry. *Case Studies of Expertise and Experience: Special Issue of Studies in History and Philosophy of Science*, 38(4): 657–666. DOI: https://doi.org/10.1016/j.shpsa.2007.09.003.

Collins, Harry, Evans, Robert and Gorman, Michael 2019 Trading zones revisited. In: Caudill, David et al. *The Third Wave in Science and Technology Studies: Future Research Directions on Expertise and Experience.* Palgrave-MacMillan. pp. 275–281. DOI: https://doi.org/10.1007/978-3-030-14335 -0_15.

Collins, Harry, Evans, Robert, Ribeiro, Rodrigo and Hall, Martin 2006 Experiments with interactional expertise. *Studies in History and Philosophy of Science*, 37(A/4): 656–674. DOI: https://doi.org/10.1016/j.shpsa.2006.09.005.

Collins, Harry, Evans, Robert and Weinel, Martin 2017 STS as science or politics? *Social Studies of Science*, 47(4): 580–586. DOI: http://dx.doi.org /10.1177/0306312717710131. PMid: 28639540.

Collins, Harry, Evans, Robert, Weinel, Martin, Lyttleton-Smith, Jennifer, Bartlett, Andrew and Hall, Martin 2017 The Imitation Game and the nature of mixed methods. *Journal of Mixed Methods Research*, 11(4): 510–527. DOI: http://dx.doi.org/10.1177/1558689815619824.

Collins, Harry and Harrison, Robert G 1975 Building a TEA laser: The caprices of communication. *Social Studies of Science*, 5(4): 441–450. DOI: https://doi.org/10.1177/030631277500500404.

Collins, Harry and Kusch, Martin 1998 *The shape of actions: What humans and machines can do.* MIT Press. DOI: https://doi.org/10.7551/mitpress /6200.001.0001.

Collins, Harry, Leonard-Clarke, Willow and Mason-Wilkes, Will Forthcoming, 2022 Scientific conferences, socialisation and lockdown: A conceptual and empirical enquiry. *Social Studies of Science*.

Collins, Harry, Leonard-Clarke, Willow and O'Mahoney, Hannah 2019 "Um, Er": How meaning varies between speech and its typed transcript. *Qualitative Research*, 19(6): 653–668. DOI: https://doi.org/10.1177/14687941188 16615.

Collins, Harry and Pinch, Trevor 1979 The construction of the paranormal: Nothing unscientific is happening. In: Wallis, Roy. *Sociological Review Monograph. No. 27: On the Margins of Science: The Social Construction of Rejected Knowledge.* Keele University Press. pp. 237–270. DOI: https://doi.org/10.1111 /j.1467-954X.1979.tb00064.x

Collins, Harry and Pinch, Trevor 2005 *Dr Golem: How to think about medicine.* University of Chicago Press. DOI: https://doi.org/10.7208/chicago /9780226113692.001.0001.

Collins, Harry and Reber, Arthur 2013 Ships that pass in the night. *Philosophia Scientiae*, 17(3): 135–154. DOI: https://doi.org/10.4000/philosophias cientiae.893.

Collins, Harry and Sanders, Gary 2007 They give you the keys and say "Drive It": Managers, referred expertise, and other expertises. In: Collins, Harry. *Case Studies of Expertise and Experience: Special Issue of Studies in History and Philosophy of Science*, 38(4): 621–641. DOI: https://doi.org/10.1016/j .shpsa.2007.09.002.

Collins, Harry, Shrager, Jeff, Conley, Shannon, Hale, Rachel and Evans, Robert Forthcoming, 2023 Hyper-normal science: Its significance for the future of scientific conferences, the replication crisis and the history of science studies. *Perspectives on Science*.

Collins, Harry, Weinel, Martin and Evans, Robert 2010 The politics and policy of the third wave: New technologies and society. *Critical Policy Studies*, 4(2): 185–201. DOI: https://doi.org/10.1080/19460171.2010.490642.

Collins, Randall 1998 *The sociology of philosophies: A global theory of intellectual change.* The Belknap Press of Harvard University Press.

Collins, Randall 2004 *Interaction ritual chains.* Princeton University Press. DOI: https://doi.org/10.1515/9781400851744.

Coppedge, Michael, Gerring, John, Glynn, Adam, Knutsen, Carl Henrik. Lindberg, Staffan, Pemstein, Daniel, Seim, Brigitte, Skaaning, Svend-Erik and Teorell, Jan 2020 *Varieties of democracy: Measuring two centuries of political change.* Cambridge University Press. DOI: https://doi.org /10.1017/9781108347860.

Corning, Peter A 1982 Durkheim and Spencer. *British Journal of Sociology*, 33(3): 359–382. DOI: https://doi.org/10.2307/589482.

Coulter, Keith and Roggeveen, Anne 2012 Like it or not: Consumer responses to word-of-mouth communication in on-line social networks. *Management Research Review*, 35(9): 878–899. DOI: https://doi.org/10.1108/0140917 1211256587.

Dahl, Robert 1971 *Polyarchy: Participation and opposition.* Yale University Press.

Davis, Wayne 2019 Implicature. In: Zalta, Edward N. *The Stanford Encyclopedia of Philosophy* (Fall 2019 Edition). Available at: https://plato.stanford.edu /archives/fall2019/entries/implicature/ [accessed 09/12/2021].

Dawson, Andrew and Innes, Martin 2019 How Russia's Internet Research Agency built its disinformation campaign. *Political Quarterly*, 90(2): 245–257. DOI: https://doi.org/10.1111/1467-923X.12690.

Delamont, Sara 1989 *Knowledgeable women: Structuralism and the reproduction of elites.* Routledge.

Digital Culture, Media and Sport Parliamentary Select Committee 2019 *Disinformation and fake news: Final report.* House of Commons. Available at: https://committees.parliament.uk/committee/378/digital-culture-media -and-sport-committee/news/103668/disinformation-and-fake-news-final -report-published/ [accessed 09/12/2021].

Disabled Student Sector Leadership Group 2017 *Inclusive teaching and learning in Higher Education.* Department for Education. Available at: https:// www.gov.uk/government/publications/inclusive-teaching-and-learning-in -higher-education [accessed 08/12/2021].

Dunbar, Robin I M 2012 Social cognition on the internet: Testing constraints on social network size. *Philosophical Transactions of the Royal Society B*, 367: 2192–2201. DOI: https://doi.org/10.1098/rstb.2012.0121. PMid: 22734062. PMCid: PMC3385686.

Durant, Darrin 2018 Servant or partner? The role of expertise and knowledge in democracy. *The Conversation*, 9 March. https://theconversation.com /servant-or-partner-the-role-of-expertise-and-knowledge-in-democracy -92026 [accessed 08/12/2021].

Durant, Darrin 2019 Ignoring experts. In: Caudill, David et al. *The Third Wave in Science and Technology Studies: Future Research Directions on Expertise and Experience.* Palgrave Macmillan. pp. 33–52. DOI: https://doi .org/10.1007/978-3-030-14335-0_3.

Earl, J and Kimport, K 2011 *Digitally enabled social change.* MIT Press. DOI: https://doi.org/10.7551/mitpress/9780262015103.001.0001.

Eckman, Paul 2009 *Telling lies: Clues to deceit in the marketplace, politics, and marriage.* W W Norton.

Eliasoph, Nina 1998 *Avoiding politics: How Americans produce apathy in everyday life.* Cambridge University Press. DOI: https://doi.org/10.1017/CBO 9780511583391.

Etzioni, Amitai 1988 *The moral dimension: Toward a new economics.* Free Press.

Evans, Robert 2011 Collective epistemology: The intersection of group membership and expertise. In: Schmid, H B et al. *Collective epistemology.* Ontos Verlag. pp. 177–202. DOI: https://doi.org/10.1515/9783110322583.177.

Evans, Robert 2021 SAGE advice and political decision-making: 'Following the science' in times of epistemic uncertainty. *Social Studies of Science.* DOI: http://dx.doi.org/10.1177/03063127211062586. PMid: 34963397.

Evans, Robert, Collins, Harry, Weinel, Martin, Lyttleton-Smith, Jennifer, O'Mahoney, Hannah and Leonard-Clarke, Willow 2019 Groups and individuals: Conformity and diversity in the performance of gendered identities. *The British Journal of Sociology,* 70(4): 1561–1581. DOI: http://dx.doi.org /10.1111/1468-4446.12507. PMid: 30351452.

Evans, Robert and Crocker, Helen 2013 The Imitation Game as a method for exploring knowledge(s) of chronic illness. *Methodological Innovations Online,* 8(1): 34–52. DOI: http://dx.doi.org/10.4256/mio.2013.003.

Evans, Robert and Kotchetkova, Inna 2009 Qualitative research and deliberative methods: Promise or peril? *Qualitative Research,* 9(5): 625–643. DOI: http://dx.doi.org/10.1177/1468794109343630.

Evans, Robert and Plows, Alexandra 2007 Listening without prejudice?: Rediscovering the value of the disinterested citizen. *Social Studies of Science,* 37(6): 827–853. DOI: https://doi.org/10.1177/0306312707076602.

Festinger, Leon, Riecken, Henry W and Schachter, Stanley 1956 *When prophecy fails.* Harper. DOI: https://doi.org/10.1037/10030-000.

Fine, Gary A 2021 *The hinge: Civil society, group cultures, and the power of local commitments.* University of Chicago Press. DOI: https://doi.org/10.7208 /chicago/9780226745831.001.0001.

Fisher, Lucy 2020 Russia accused of trying to hack vaccine research. *The Times,* 17 July. https://www.thetimes.co.uk/article/russian-actors-tried-to-influence -2019-general-election-says-dominic-raab-z57j6s825 [accessed 08/12/2021].

Fleck, John 2016 *Water is for fighting over: And other myths about water in the West.* Island Press.

Fleck, Ludwik 1979 *Genesis and development of a scientific fact.* University of Chicago Press (first published in German in 1935 as *Entstehung und*

Entwicklung einer wissenschaftlichen Tatsache: Einführung in die Lehre vom Denkstil und Denkkollektiv).

Fu, Pei-Wen, Wu, Chi-Cheng and Cho, Yung-Jan 2017 What makes users share content on Facebook? Compatibility among psychological incentive, social capital focus and content type. *Computers in Human Behavior,* 67(Feb): 23–32. DOI: https://doi.org/10.1016/j.chb.2016.10.010.

Gane, Nicholas 2001 Zygmunt Bauman: Liquid modernity and beyond. *Acta Sociologica,* 44(3): 267–275. DOI: http://dx.doi.org/10.1177/0001699 30104400306.

Giddens, Anthony 1990 *The consequences of modernity.* Stanford University Press.

Giddens, Anthony 1994 Living in a post-traditional society. In: Beck, Ulrich et al. *Reflexive Modernization: Politics, Tradition and Aesthetics in the Modern Social Order.* Polity Press. pp. 56–109.

Giles, Jim 2006 Sociologist fools physics judges. *Nature* 442: 8. DOI: https://doi.org/10.1038/442008a. PMid: 16823419.

Goffman, Erving 1959 *The presentation of self in everyday life.* Penguin.

Goffman, Erving 1961 *Asylums; Essays on the condition of the social situation of mental patients and other inmates.* Penguin.

Goffman, Erving 1967 *Interaction ritual: Essays on face-to-face behaviour.* Oxford University Press.

Goffman, Erving 1983 The interaction order. *American Sociological Review,* 48(1): 1–17. DOI: https://doi.org/10.2307/2095141.

Granovetter, Mark 1973 The strength of weak ties. *American Journal of Sociology,* 78: 1360–1380. DOI: https://doi.org/10.1086/225469.

Granovetter, Mark 1985 Economic action and social structure: The problem of embeddedness. *American Sociological Review,* 91(3): 481–510. DOI: https://doi.org/10.1086/228311.

Grene, Marjorie (ed.) 1969 *Knowing and being: Essays by Michael Polanyi.* Routledge & Kegan Paul.

Griffith, George V, Phelps, Elizabeth Stuart and Eliot, George 2001 An epistolatory friendship: The letters of Elizabeth Stuart Phelps to George Eliot. *Legacy,* 18(1): 94–100. Project MUSE. DOI: https://doi.org/10.1353/leg .2001.0002.

Hall, S, Hobson, D, Lowe, A and Willis, P 2003 [1980] *Culture, media, language: Working papers in cultural studies, 1972–79.* Taylor & Francis. DOI: https://doi.org/10.4324/9780203381182.

Hamilton, W 1973 *Plato, The Phaedrus, translated by Walter Hamilton.* Penguin.

Hargreaves, Ian, Lewis, Justin and Speers, Tammy 2003 *Towards a better map: Science, the public and the media.* ESRC.

Hartshorne, Joshua K, Tenenbaum, Joshua B, Pinker, Steven 2018 A critical period for second language acquisition: Evidence from 2/3 million English speakers. *Cognition,* 177:263–77. DOI:https://dx.doi.org/10.1016/j.cognition .2018.04.007. PMid: 29729947. PMCid: PMC6559801.

Hasher, Lynn, Goldstein, David and Toppino, Thomas 1977 Frequency and the conference of referential validity. *Journal of Verbal Learning and Verbal Behaviour*, 16(1): 107–112. DOI: https://doi.org/10.1016/S0022-5371 (77)80012-1.

Hayek, Friedrich A 1944 *The road to serfdom*. Routledge.

Healy, Kieran 2017 Fuck nuance. *Sociological Theory*, 35(2): 118–127. DOI: https://doi.org/10.1177/0735275117709046.

Hill, Jane H 2008 *The everyday language of white racism*. Wiley. DOI: https://doi.org/10.1002/9781444304732.

Hirsch, J 2015 Elon Musk's growing empire is fueled by $4.9 billion in government subsidies. *Los Angeles Times*, 30 May. Available at: http://www.latimes.com/business/la-fi-hy-musk-subsidies-20150531-story.html [accessed 08/12/2021].

Innes, Martin, Dobreva, Diyana and Innes, Helen 2021 Disinformation and digital influencing after terrorism: Spoofing, truthing and social proofing. *Contemporary Social Science*, 16(2): 241–255. DOI: http://dx.doi.org/10.10 80/21582041.2019.1569714.

Jaakonmäki, Roope, Müller, Oliver and von Brocke, Jan 2017 The impact of content, context and creator on user engagement in social media marketing. In: Proceedings of the 50th Hawaii International Conference on System Sciences Jan 2017. Available at: http://hdl.handle.net/10125/41289 [accessed 09/12/2021]. DOI: https://doi.org/10.24251/HICSS.2017.136.

Jasanoff, Sheila and Simmet, Hilton R 2017 No funeral bells: Public reason in a 'post-truth' age. *Social Studies of Science*, 47(5): 751–770. DOI: https://doi.org/10.1177/0306312717731936. PMid: 29034796.

Jowett, Gareth S and O'Donnell, Victoria 2018 *Propaganda & persuasion* (7th edition). SAGE.

Kaiser, Brittany 2019 *Targeted: The Cambridge Analytica whistleblower's inside story of how Big Data, Trump, and Facebook broke democracy and how it can happen again*. Harper Collins.

Karnad, Arun 2013 *Student use of recorded lectures: A report reviewing recent research into the use of lecture capture technology in Higher Education, and its impact on teaching methods and attendance*. London School of Economics. Available at: http://eprints.lse.ac.uk/50929/ [accessed 09/12/2021].

Kata, Anna 2012 Anti-vaccine activists, Web 2.0, and the postmodern paradigm – An overview of tactics and tropes used online by the anti-vaccination movement. *Vaccine*, 30(25): 3778–3789. DOI: https://doi.org/10.1016/j.vaccine.2011.11.112. PMid: 22172504.

Keating, Brian 2018 *Losing the Nobel Prize: A story of cosmology, ambition, and the perils of science's highest honor*. W W Norton.

Kennedy, Jonathan 2019 Populist politics and vaccine hesitancy in Western Europe: An analysis of national-level data. *The European Journal of Public Health*, 29(3): 512–516. DOI: https://doi.org/10.1093/eurpub/ckz004. PMid: 30801109.

Kitzinger, J 2004 *Framing abuse: Media influence and public understanding of sexual violence against children*. Pluto Books.

Kuhl, Patricia K 2004 Early language acquisition: Cracking the speech code. *Nature Reviews Neuroscience,* 5: 831–843. DOI: https://doi.org/10.1038 /nrn1533. PMid: 15496861.

Kuhl, Patricia K 2010 Brain mechanisms in early language acquisition. *Neuron,* 67(5): 713–727. DOI: https://dx.doi.org/10.1016/j.neuron.2010.08.038. PMid: 20826304. PMCid: PMC2947444.

Kuhn, Thomas S 1959 The essential tension: Tradition and innovation in scientific research. In: Taylor, C W. *The Third University of Utah Research Conference on the Identification of Scientific Talent.* University of Utah Press. pp. 162–174.

Kuhn, Thomas S 1977 *The essential tension: Selected studies in scientific tradition and change.* University of Chicago Press. DOI: https://doi.org/10.7208 /chicago/9780226217239.001.0001.

Labinger, Jay and Collins, Harry (eds.) 2001 *The one culture?: A conversation about science.* University of Chicago Press. DOI: https://doi.org/10.7208 /chicago/9780226467245.001.0001.

Lakoff, Robin 1973 Language and woman's place. *Language in Society* 2(1): 45–79. DOI: http://dx.doi.org/10.1017/S0047404500000051.

Lakoff, Robin 1975 *Language and woman's place.* Harper & Row.

Lepore, Jill 2019 Taking history personally: Jill Lepore considers the rise of the conspiracy theory. *The Times Literary Supplement,* 6 August. Available at: https://www.the-tls.co.uk/articles/public/modernity-conspiracy-theory -jill-lepore/ [accessed: 02/09/2019].

Lewis, Michael 2010 *The big short: Inside the doomsday machine.* W W Norton.

Lewis, Michael 2014 *Flash boys: A Wall Street revolt.* W W Norton.

Lewis, Rebecca 2018 *Alternative influence: Broadcasting the reactionary right on YouTube.* Data & Society Research Institute. Available at: https://datasociety.net /wp-content/uploads/2018/09/DS_Alternative_Influence.pdf [accessed:21/01 /2020].

Luhmann, Niklas 1998 Familiarity, confidence, trust: Problems and alternatives. In: Gambetta, D. *Trust: Making and Breaking Cooperative Relations.* Blackwell. pp. 94–107.

Lynch, Michael 2017 STS, symmetry and post-truth. *Social Studies of Science,* 47(4): 593–599. DOI: https://doi.org/10.1177/0306312717720308. PMid: 28791930.

MacKenzie, Donald 2019 Just how fast? *London Review of Books.* https://www .lrb.co.uk/v41/n05/donald-mackenzie/just-how-fast [accessed 08/12/2021].

Madeddhu, Paolo 2020 Preventing a covid-19 pandemic. *British Medical Journal,* 12 March. DOI: https://doi.org/10.1136/bmj.m810. PMid: 32111649.

Malhotra, Arvind, Majchrzak, Ann, Carman, Robert and Lott, Vern 2001 Radical innovation without collocation: A case study at Boeing-Rocketdyne. *Management Information Systems Quarterly,* 25(2): 229–249. DOI: https://doi.org/10.2307/3250930.

Manstead, Anthony S R, Lea, Martin and Goh, Jeannine 2011 Facing the future: Emotion communication and the presence of others in the age of video-mediated communication. In: Kappas, Arvid and Kramer, Nicole C. *Face-to-Face Communication Over the Internet: Emotions in a Web of Culture, Language and Technology.* Cambridge University Press. pp. 144–176. DOI: https://doi.org/10.1017/CBO9780511977589.009.

Margetts, Helen, John, Peter, Hale, Scott and Yasseri, Taha 2016 *Political turbulence: How social media shape collective action.* Princeton University Press. DOI: https://doi.org/10.2307/j.ctvc773c7.

Marwick, Alice 2013 *Status update: Celebrity, publicity and branding in the social media age.* Yale University Press.

Mason-Wilkes, Will 2018 *Science as religion? Science communication and Elective Modernism.* Thesis (PhD), Cardiff University. http://orca.cf.ac.uk/109735/.

Mauranen, Anna 2003 "But here's a flawed argument": Socialization into and through metadiscourse. In: Leistyna, Pepi and Myer, Charles F. *Corpus Analysis: Language Structure and Language Use.* Rodopi. pp. 19–34. DOI: https://doi.org/10.1163/9789004334410_003.

Mazzucato, Mariana 2013 *The entrepreneurial state: Debunking the public vs. private myth in risk and innovation.* Anthem Press.

Mazzucato, Mariana 2017 Mission-oriented innovation policy: Challenges and opportunities. RSA (Royal Society for the encouragement of Arts, Manufactures and Commerce) and UCL. Available at: https://www.thersa.org/globalassets/pdfs/reports/mission-oriented-policy-innovation-report.pdf [accessed 08/12/2021].

McLeod, Carolyn 2021 Trust. In: Zalta, Edward N. *The Stanford Encyclopedia of Philosophy* (Fall 2021 Edition). Available at: https://plato.stanford.edu/archives/fall2021/entries/trust/ [accessed 09/12/2021].

McPherson, Miller, Smith-Lovin, Lynn and Cook, James M 2002 Birds of a feather: Homophily in social networks. *Annual Review of Sociology*, 27: 415–444. DOI: http://dx.doi.org/10.1146/annurev.soc.27.1.415.

McQuail, Denis 2005 *McQuail's mass communication theory* (5th edition). SAGE Publications.

Mehrabian, Albert 1972 *Nonverbal communication.* Aldine-Atherton.

Merton, Robert K 1942 Science and technology in a democratic order. *Journal of Legal and Political Sociology*, 1(1): 115–126.

Michels, Robert 1911 *Political parties: A sociological study of the oligarchical tendencies of modern democracy* [Transaction Press edition, 1966].

Mirowski, Philip 2018 The future of open science. *Social Studies of Science*, 48(2): 171–203. DOI: https://doi.org/10.1177/0306312718772086. PMid: 29726809.

Mirowski, Philip 2019 Hell is truth seen too late. *boundary 2*, 46(1): 1–53. DOI: https://doi.org/10.1215/01903659-7271327.

Mirowski, Philip 2020 Democracy expertise and the post truth era: An inquiry into the contemporary politics of STS. Available at: https://www.academia

.edu/42682483/Democracy_Expertise_and_the_Post_Truth_Era_An _Inquiry_into_the_Contemporary_Politics_of_STS [accessed 09/12/2021].

Mirowski, Philip (forthcoming) STS, platform capitalism and the conundrum of expertise. *Circus Bazar Magazine.*

MNP 2017 A review of the 2016 Horse River wildfire: Alberta Agriculture and Forestry preparedness and response. MNP LLP. Available at: https://www .alberta.ca/assets/documents/Wildfire-MNP-Report.pdf [accessed 09/12/ 2021].

Monbiot, George 2016 Neoliberalism – The ideology at the root of all our problems. *The Guardian*, 15 April. Available at: https://www.theguardian .com/books/2016/apr/15/neoliberalism-ideology-problem-george-monbiot [accessed 08/12/2021].

Mouffe, Chantal 2000 *The democratic paradox.* Verso.

Mueller, Robert 2019 *Report on the investigation into Russian interference in the 2016 presidential election.* US Department of Justice.

Müller, Jan-Werner 2017 *What is populism?* Penguin Books. DOI: https://doi .org/10.9783/9780812293784.

Muñoz, Caroline and Towner, Terri 2017 The image is the message: Instagram marketing and the 2016 presidential primary season. *Journal of Political Marketing*, 16(3–4): 290–318. DOI: https://doi.org/10.1080/15377857.2 017.1334254.

Nahai, Nathalie 2017 *Webs of influence: The psychology of online persuasion.* Pearson: New York.

Nonaka, Ikujiro and Takeuchi, Hirotaka 1995 *The knowledge-creating company.* Oxford University Press.

Nordmann, Emily and McGeorge, Peter 2018 Lecture capture in higher education: Time to learn from the learners. *PsyArXiv*, May 2018. DOI: https:// doi.org/10.31234/osf.io/ux29v.

Norenzayan, Ara 2014 *Big gods: How religion transformed cooperation and conflict.* Princeton University Press. DOI: https://doi.org/10.1515/9781400 848324.

Novinite 2020 Not fake news: Ibuprofen and cortisone may worsen your condition if you are infected with COVID-19 – UPDATED. *Novinite.com*, 15 March. Available at: https://www.novinite.com/articles/203622/Fake+New s:+Ibuprofen+and+Cortisone+may+Worsen+your+Condition+if+you+ar e+Infected+with+COVID-19 [accessed 09/12/2021].

Open Society Foundations 2021 *What we do.* Available at: https://www.open societyfoundations.org/what-we-do [accessed 08/12/2021].

Oreskes, Naomi and Conway, Erik M 2010 *Merchants of doubt: How a handful of scientists obscured the truth on issues from tobacco smoke to global warming.* Bloomsbury Press.

Packard, Vance 1957 *The hidden persuaders.* D McKay Co.

Peters, Diane 2020 Women academics worry the pandemic is squeezing their research productivity. *University Affairs* 7 July. Available at: https://www

.universityaffairs.ca/news/news-article/women-academics-worry-the-pan
demic-is-squeezing-their-research-productivity [accessed 08/12/2021].

Pitkin, Hanna 1967 *The concept of representation*. University of Los Angeles Press.

Polanyi, Karl 1944 *The great transformation: The political and economic origins of our time*. Beacon Press.

Polanyi, Michael 1966 The logic of tacit inference. *Philosophy*, 41(155): 1–18. Available at: https://www.jstor.org/stable/3749034. DOI: https://doi.org /10.1017/S0031819100066110.

Pomerantsev, Peter 2019a *This is not propaganda: Adventures in the war against reality*. Faber and Faber Ltd.

Pomerantsev, Peter 2019b Control, shift, delete. *Guardian Review*, 27 July 2019. pp. 7–11.

Popper, Karl R 1959 *The logic of scientific discovery*. Harper & Row. DOI: https://doi.org/10.1063/1.3060577.

Purvanova, Radostina K 2014 Face-to-face versus virtual teams: What have we really learned? *The Psychologist-Manager Journal*, 17(1): 2–29. DOI: https://doi.org/10.1037/mgr0000009.

Quality and Assurance Agency for Higher Education (QAA) 2019 *Quality and credit frameworks, Annex D outcome classification descriptions for FHEQ Level 6 and FQHEIS Level 10 degrees*. Available at: https://www.qaa .ac.uk//en/quality-code/qualifications-frameworks [accessed 08/12/2021].

Ramjug, Peter 2020 Think everyone will be clamoring to get a COVID-19 vaccine? Think again, study says. *MedicalXpress*, 6 August. Available at: https://medicalxpress.com/news/2020-08-clamoring-covid-vaccine.html [accessed 09/12/2021].

Reber, Arthur S 2011 An epitaph for grammar. In: Sanz, Cristina and Loew, Ronald P. *Implicit and Explicit Language Learning*. Georgetown University Press. pp. 23–34.

Reiner, Robert 1992 *The politics of the police* (2nd edition). Harvester Wheatsheaf.

Ribeiro, Rodrigo and Collins, Harry 2007 The bread-making machine, tacit knowledge and two types of action. *Organization Studies*, 28(9): 1417–1433. DOI: https://doi.org/10.1177/0170840607082228.

Rid, Thomas 2020 *Active measures: The secret history of disinformation and political warfare*. Farrar, Strauss & Giroux.

Ryan-Collins, Josh 2018 *Why can't you afford a home?* Polity Press.

Ryan-Collins, Josh, Greenham, Tony, Werner, Richard and Jackson, Andrew 2012 *Where does money come from? A guide to the UK monetary system* (2nd edition). New Economics Foundation.

Ryfe, Davide Michael 1999 Franklin Roosevelt and the fireside chats. *Journal of Communication*, 49(4): 80–103. DOI: http://dx.doi.org/10.1111/j.1460 -2466.1999.tb02818.x.

Schutz, Alfred 1972 *The phenomenology of the social world*. Northwestern University Press.

Shapin, Steven 1994 *A social history of truth: Civility and science in seventeenth-century England*. University of Chicago Press. DOI: https://doi.org/10.7208 /chicago/9780226148847.001.0001. PMid: 10135275.

Shapiro, Susan P 1987 The social control of interpersonal trust. *American Journal of Sociology*, 93(3): 623–658. DOI: https://doi.org/10.1086/228791.

Shapiro, Susan P 2005 Agency theory. *Annual Review of Sociology*, 31(1): 263–284. DOI:http://dx.doi.org/10.1146/annurev.soc.31.041304.122159.

Shapiro, Susan P 2012 The grammar of trust. In: Pixley, Jocelyn. *New Perspectives on Emotions in Finance: The Sociology of Confidence, Fear and Betrayal*. Routledge. pp. 99–234.

Sismondo, Sergio 2017a Post-truth? *Social Studies of Science*, 47(1): 3–6. DOI: https://doi.org/10.1177/0306312717692076. PMid: 28195024.

Sismondo, Sergio 2017b Casting a wider net: A reply to Collins, Evans and Weinel. *Social Studies of Science*, 47(4): 587–592. DOI: https://doi.org /10.1177/0306312717721410. PMid: 28791928.

Smith, Edward Bishop, Brands, Raina A, Brashears, Matthew E and Kleinbaum, Adam M 2020 Social networks and cognition. *Annual Review of Sociology*, 46(1): 159–174. DOI: https://doi.org/10.1146/annurev-soc-121919 -054736.

Speer, Susan A and Stokoe, Elizabeth (eds.) 2011 *Conversation and gender*. Cambridge University Press.

Stevenson, Charles L 1938 Persuasive definitions. *Mind*, 47(187): 331–350. DOI: https://doi.org/10.1093/mind/XLVII.187.331.

Stouffer, Samuel A 1949 *The American soldier*. Princeton University Press.

Susskind, Lawrence and Field, Patrick 1996 *Dealing with an angry public: The mutual gains approach to resolving disputes*. Free Press.

Swinger, Nathaniel, De-Arteaga, Maria, Heffernan, Neil Thomas, Leiserson, Mark D M and Kalai, Adam Tauman 2019 What are the biases in my word embedding? Available at: https://arxiv.org/pdf/1812.08769.pdf.DOI: https:// doi.org/10.1145/3306618.3314270.

The Linux Kernel (no date) How the development process works. Available at: https://www.kernel.org/doc/html/v4.15/process/2.Process.html [accessed 09/12/2021].

Thorpe, Charles 2016 *Necroculture*. Palgrave Macmillan. DOI: https://doi.org /10.1057/978-1-137-58303-1.

Toynbee, Polly 2020 The Daily Mail has turned against the anti-vaxxers it used to champion. *The Guardian*, 13 November. https://www.theguardian.com /commentisfree/2020/nov/13/daily-mail-anti-vaxxers-paper-covid -vaccine-mmr [accessed 08/12/2021].

Tufecki, Zeynep 2017 *Twitter and tear gas: The power and fragility of networked protest*. Yale University Press.

Turing, Alan M 1950 Computing machinery and intelligence *Mind*, LIX(236): 433–460. DOI: https://doi.org/10.1093/mind/LIX.236.433.

Turkle, Sherry 2011 *Alone together*. Hachette.

Turkle, Sherry 2015 *Reclaiming conversation: The power of talk in a digital age.* Penguin Random House.

Veirman, Marijke, Cauberghe, Veroline and Hudders, Liselot 2017 Marketing through Instagram influencers: The impact of number of followers and product divergence on brand attitude. *International Journal of Advertising,* 36(5): 798–828. DOI: https://doi.org/10.1080/02650487.2017.1348035.

Vieira, Helena 2019 The system is not immune to capture by interest groups. London School of Economics blog. Available at: https://blogs.lse.ac.uk/businessreview/2019/06/04/blockchain-governance-the-system-is-not-immune-to-capture-by-interest-groups/ [accessed 09/12/2021].

Wakefield, Andrew J, Murch, Simon H, Anthony, Andrew et al. 1998 Ileal-lymphoid-nodular hyperplasia, non-specific colitis, and pervasive developmental disorder in children. *The Lancet,* 351(9103): 637–641. DOI: https://doi.org/10.1016/S0140-6736(97)11096-0.

Wehrens, Rik and Walters, Bethany Hipple 2017 Understanding each other in the medical encounter: Exploring therapists' and patients' understanding of each other's experiential knowledge through the Imitation Game. *Health,* 22(6): 558–579. DOI: http://dx.doi.org/10.1177/1363459317721100. PMid: 28770633. PMCid: PMC6168741.

Wikipedia 2021a Albert Mehrabian. Available at: https://en.wikipedia.org/wiki/Albert_Mehrabian [accessed 09/12/2021].

Wikipedia 2021b 2011–2013 Russian protests. Available at: https://en.wikipedia.org/wiki/2011–2013_Russian_protests [accessed 09/12/2021].

Willsher, Kim 2020 Anti-inflammatories may aggravate Covid-19, France advises. *The Guardian,* 14 March. https://www.theguardian.com/world/2020/mar/14/anti-inflammatory-drugs-may-aggravate-coronavirus-infection [accessed 09/12/2021].

Wilson, Bryan (ed.) 1970 *Rationality.* Blackwell.

Witthaus, Gabi, and Robinson, Carol 2015 Lecture capture literature review: A review of the literature from 2012–2015. *Centre for Academic Practice, Loughborough University.* Available at: https://repository.lboro.ac.uk/articles/Lecture_capture_literature_review_A_review_of_the_literature_from_2012-2015/9368876 [accessed 09/12/2021].

Wylie, Christopher 2019 *Mindf*cked: Cambridge Analytica and the plot to break America.* Random House.

Wykstra, Stephanie 2016 Can robots help resolve the reproducibility crisis? *Slate,* 30 June. Available at: https://slate.com/technology/2016/06/automating-lab-research-could-help-resolve-the-reproducibility-crisis.html [accessed 08/12/2021].

Zuboff, Shoshana 2019 *The age of surveillance capitalism: The fight for a human future at the new frontier of power.* Profile Books.

Index

9 781911 653295